Weather at Sea

Weather
at
Sea

The Seamanship Series

David Houghton and Fred Sanders

International Marine Publishing Company
Camden, Maine 04843

© 1988 Highmark Publishing Ltd.

Published by International Marine Publishing Co., a division of Highmark Publishing Ltd., Camden, Maine 04843.

Adapted from the 1986 edition published in the United Kingdom by Fernhurst Books.

Typset by Camden Type 'n Graphics, Camden, ME.
Printed and bound by BookCrafters, Chelsea, MI.
Designed by Ken Gross
Production by Janet Robbins
Edited by Jonathan Eaton

10 9 8 7 6 5 4 3 2 1

Library of Congress Cataloging-in-Publication Data

Sanders, Frederick, 1923–
    Weather at sea / Frederick Sanders, David Houghton.
        p.   cm. — (The Seamanship series)
    Includes index.
    ISBN 0-87742-959-6 : $12.95
    1. Meteorology, Maritime.   I. Houghton, David D.   II. Title
III. Series.
QC994.S25   1988   87-35346
551.5'09162—dc19   CIP

Contents

Preface and Acknowledgments

This book offers a more challenging explanation of weather systems than is usually found in a volume written for the non-meteorologist. It reflects our great increase over the last two or three decades in understanding how systems form and work. You can be flexible in going back and forth between the later chapters presenting rules and observations with which you may already be somewhat familiar and the earlier chapters which put forward a contemporary basis of understanding and will probably be newer to you.

We hope you will come to appreciate that the weather is not as fickle as it sometimes appears. There is an ordered sequence in the unfolding of a thunderstorm or the development and oscillations of a sea breeze. A fundamental understanding of the mechanisms will make you a better local forecaster—more prepared for the trials of weather, and more often on the favored side of the course when the fleet beats to the weather mark. We hope too that this book will lead you to share with those of us who are both meteorologists and sailors an admiration for the structures, a respect for the hazards, and an enjoyment of the beauties which the atmosphere presents.

We wish to express our thanks to David Burch of the Starpath School of Navigation in Seattle, Washington for reviewing the manuscript and offering valuable suggestions from the West Coast perspective. We are grateful also to Andy Horvitz of NOAA in Washington, D.C., for help in selection of satellite pictures.

Introduction—How to Use This Book

As in celestial navigation, diesel engine repair, and a host of other subjects one could name, there are two ways to approach the study of marine weather. The first is to seek a working knowledge—the ability to recognize good or bad portents in clouds, barometric readings, temperature changes, and weather maps, and to act accordingly. The second goes beyond the first to ask why weather happens as it does. Why do low- and high-pressure systems form, why do they move, and what causes their characteristic weather patterns? What shapes the wind? How do thunderstorms and tropical storms arise? How do the jet streams and weather aloft interact with weather at the sea surface?

Chapters 2 through 5 in this book present the why's of weather, the mechanisms that answer the questions posed above. They are perhaps more challenging than other books on marine weather because they reflect the great increase in understanding of the weather gained in the last 25 years, but for the same reason they are in the end, we think, more accurate and sound. These chapters are included because we believe that the deeper one's understanding of a subject, the more lasting and satisfying it will be. Nevertheless, we do not want you to get bogged down in these chapters, because what you will learn in them, although useful, is not essential for extracting the working knowledge presented in the rest of the book. If some statements are unclear, continue reading and go back to them later. Chapter 6 will put you on more familiar ground and Chapters 2 through 5 will be much clearer when you finish the book and return to them. The Glossary, too, should help in this regard. We believe that your effort will be rewarded with a truly thorough and lasting understanding of weather at sea.

1

Weather Systems— A First Look

It is second-nature for a sailor to watch the wind and the sky. On a sunny day with light wind a thin veil of high cloud rises in the west as the breeze swings toward the southeast and picks up to respectable strength. Our experience tells us that tomorrow may be rainy. When the rain has ended and the wind has shifted to the northwest, we expect to see breaks appearing soon in the low cloud, and perhaps clear skies by sunset. Let's back away and look at the big picture.

We have all seen from satellite pictures on television how clouds trace out movements of the air. Let's imagine a short sequence of pictures covering, say, two or three hours, and imagine the sequence speeded up so that it moves as we watch. Assume the portion of the globe covered by the pictures is North America, although the descriptions that follow apply equally to the middle latitudes anywhere in the northern hemisphere (the region roughly from 40° N to 60° N in figure 2.2). Suppose there are no big clear areas, and suppose that the high clouds do not obscure the low ones. Then near the surface we should see counterclockwise inward-spiralling and clockwise outflowing whirls, each perhaps half the width of North America. Around one of the former whirls, which are called *cyclones*, the low clouds may move swiftly, so we know that a gale is blowing at the surface. In contrast, the winds around the clockwise whirls, called *anticyclones*, are light. Aloft we see one or more great meandering rivers in which the currents sweep from west to east through alternating wavelike clock-

3

wise and counterclockwise bends. At places and times the upper current flows swiftly in a narrow, jet-like ribbon (as in figure 4.2). This is a *jet stream*. Sometimes the bends at upper levels cut off into whirls like those that dominate the picture near the surface.

We will see in Chapter 4 that the barometric pressure is relatively low at the centers of the counterclockwise cyclonic whirls and bends, while it is high in the clockwise anticyclonic ones, both near the surface and aloft. So we speak of the surface *lows* and *highs* and the upper-level *troughs* and *ridges* in the overall westerly flow. Troughs and ridges can exist in the surface pattern as well.

When we let the picture sequence run for two or three days, we see that these whirls and bends in the middle latitudes, and their accompanying features in the pressure topography, are themselves moving along: the surface lows mainly to the northeast and the surface highs to the southeast, as the upper waves progress eastward. Near and ahead of the lows, and ahead of the upper troughs, are great masses of cloud and rain. Around and in advance of the highs and upper ridges only low clouds are seen. Large areas of clear sky may open up, depriving us of our tracers and leaving us to our imaginations. Along some surface troughs where warmer and colder air have impinged to form *fronts*, long rainbands appear. Each of these instances, as we shall see in the chapters that follow, signifies the ascent and descent of air. It is the ascent and descent of air that really makes "weather"—the variety of rain, cloud, and sunshine, changing from day to day and week to week. Our personal observations are just a small sampling of these great overall patterns.

Of course, the weather can change from hour to hour and from minute to minute as well, so this picture of the daily weather systems is only a broad one. A tiny dot in the satellite picture can blossom spontaneously in a few hours into a massive thunderstorm complex 100 miles or more across. Smaller and much weaker versions of the same process that produces these spectacular storms are responsible for the myriad dots of fair-weather cloud almost always seen over the tropical oceans. Even a major region of cloudiness in advance of a low-pressure center or trough may be organized into striations a few miles apart, separating lines of showers from intervening strips of sunshine. A smooth, thin film that you can barely see in the satellite picture may be a dense blanket of fog on the sea surface.

Three consecutive days of weather in the eastern Pacific Ocean shown in figures 1.1–1.3 provide examples of almost all these kinds of features. Let's concentrate our attention on the long frontal cloud band running northeast-southwest across the central part of the first map. It separates a mass of relatively warm air between the west coast of North America and the Hawaiian Islands from a vast region of air that becomes continuously colder to the northwest. The front lies at the leading edge of the temperature contrast in the colder air. Notice especially that a weak cyclone coincides with a slight deformation of the front near 30° north latitude and 150° west longitude. This low center in its infancy has a minimum pressure only slightly under 1016 millibars, while the flanking anticyclones to the east and west have high central pressure in excess of 1024 millibars. They are regions of scattered low clouds.

A day later (figure 1.2) the cyclone has moved rapidly northeastward and has deepened substantially to a minimum pressure of slightly less than 1000 millibars. A prominent circulation of winds around the center has developed.

The southeasterlies ahead of the cyclone represent a retreat of the colder air ahead of the advancing *warm front*. Here the barometer is falling rapidly and the frontal cloud band has developed into a broad shield. At its eastern extremity this cloud is high and thin, but nearer the low center it thickens and lowers to produce heavy rain.

Behind the cyclone, brisk northwesterlies have brought the colder air rapidly southeastward behind the *cold front*. The frontal cloud band here has narrowed to a sharp zone of heavy showers that last only a brief time before the pressure begins to rise quickly and the frontal clouds give way to scattered low cloud in the colder air.

To the right of the track of the low, mild southwesterlies blow in the *warm sector* where the barometer is falling only slowly and cloudiness is broken with occasional showers.

Winter weather patterns over the eastern Pacific Ocean. The solid lines are isobars (lines of equal pressure) at sea level, and the dashed line are isotherms (lines of equal temperatures) averaged over the layer from the surface to 18,000 feet. Arrows show the direction of the surface wind. The heavy line is a frontal system. Notice how it coincides with a major band of cloud in the satellite picture.

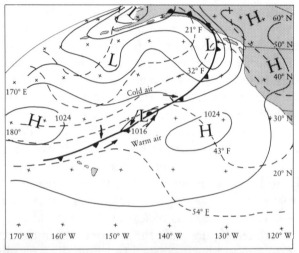

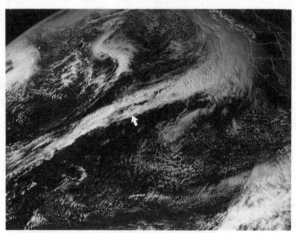

Figure 1.1. *The first day. A weak cyclone lies along a deformation of the front, denoted by the arrow in the satellite photo.*

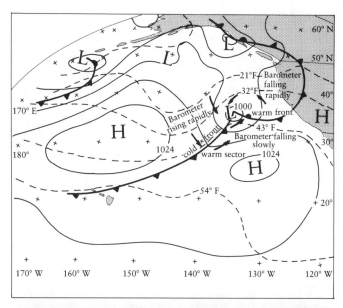

Figure 1.2. *A day later. Note the broad shield of cloud (arrow) ahead of the deepening cyclone, where colder air retreats ahead of the warm front, which advances in a counterclockwise direction as the entire system moves northeastward.*

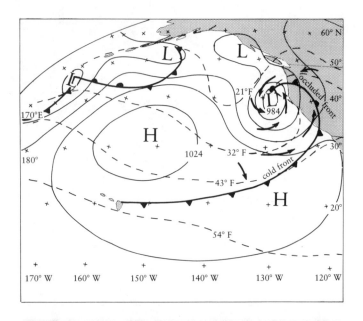

Figure 1.3. *Two days later.*

After another day (figure 1.3), the cyclone has become an intense counterclockwise vortex around which extrusions of frontal cloud and dry cloud-free colder air have become coiled. The low center, now below 984 millibars, has moved into colder air, the original warm-sector air no longer reaches the center, and the leading edge of the colder air, which has wrapped around the cyclone center and now appears in its southeastern sector, is customarily denoted an *occluded front*. The narrow cold-front cloud band is still prominent, but it is now (as before, in fact) only one of numerous banded structures in the cloud pattern before us.

Explanations customarily offered for what we see have been almost exclusively in terms of fronts. Nearly everything is viewed as the result of block-like wedges of dense cold air thrusting under or being overrun by masses of light warm air in which cloud and rain form. Cyclones are seen as arising from a wave-like deformation of a frontal line separating cold and warm air masses.

But air masses are not solid blocks. The change in temperature from the tropics to the polar latitudes is generally gradual, although in some places relatively concentrated, as we can see from the isotherms in figures 1.1–1.3. The satellite pictures show us frontal cloud bands but also many other lines and features, many of which come and go from day to day. To regard them all as the result of a multiplicity of fronts would strain our credulity.

We now understand that a cyclone requires *contrast* between cold and warm air but not a frontal *discontinuity* between them. A front is as likely to be the result of the development of a cyclone as it is to be the precursor of one. We now appreciate that the rotation of the earth about its axis, as we will see in Chapter 3, permits colder and warmer air to lie horizontally adjacent most of the time without much undercutting or overrunning, in a state of near balance between temperature and wind. The cyclones, the anticyclones, the fronts, the clouds, the rain, and the clear skies are all the result of the atmosphere's continuous process of adjustment to this balance in the face of continuously changing temperature and wind patterns. In some ways the process resembles the oscillations of the coils of a box spring as weights are continuously moved here and there on the bed.

On the finer scale meteorologists now have a good understanding of how a front forms and produces a band of rain. Other processes, however, some of them not well understood, also produce lines and various features of the cloud pattern.

How can the sailor understand and deal with this great turmoil of the earth's atmosphere, appearing at times to be almost chaotic? How can the sailor defend ship and self against the various threats the weather can pose, from gale to thunderstorm to fog bank? How can the sailor use the ordinary pleasant winds and skies to greatest advantage? To give you a modern understanding of cyclones, anticyclones, fronts and other weather systems, and to answer these practical questions, are what this book is about.

2

Weather Basics—
The Driving Forces

We stated in Chapter 1 that the ascent and descent of air create the weather we experience. Here we examine the crucial roles of heat and water vapor in these vertical movements, knowledge that we build on in Chapter 5 to understand how weather systems form. The discussion that follows is also useful to understanding the origins of cloud types (Chapter 6), hurricanes (Chapter 14), and thunderstorms (Chapter 15).

When we talk about weather we mean everything going on in the atmosphere around us: its temperature, which affects our comfort; its humidity, which not only affects how we feel but also determines how far we can see through it; its cloudiness and whether it is raining; and in particular the speed and direction in which the air is moving, which determines whether, where, and how fast we can sail.

Besides the air itself, there are two fundamental ingredients of our weather: *heat* and *water vapor*. There is also a crucial modifying factor, the *rotation* of the earth about its axis. Weather systems on the rotating earth are much different from what they would be on a static earth, even aside from the seasonal effects due to the revolution of the earth in its orbit around the sun.

Heat

Heat from the sun is the energy which, directly or indirectly, drives all the main wind systems around the world. In its simplest form it acts like a fire heating the air, which rises and spreads out, to be replaced by colder air moving in around the sides (figure 2.1). On the global scale we have heated air rising over equatorial areas and being replaced by colder air moving in from the polar regions, and on the local scale we have sea breezes when air warmed over the land rises to be replaced by colder air moving in from the sea.

Because of the rotation of the earth, moving air is continually deflected to the right of its path in the northern hemisphere and to the left in the southern hemisphere. The force that causes this deflection is known as the Coriolis force. The consequences of this force are easily seen in the trade wind regions (figure 2.2), where the deflection of air moving toward the equator makes the winds northeasterly in the northern hemisphere and southeasterly in the southern hemisphere.

The farther you are from the equator the stronger is the Coriolis deflecting force. As the latitude increases beyond about 30 degrees both north and south—that is, poleward of the trade wind regions—the Coriolis deflecting force becomes so strong that the simple trade wind systems are replaced by something more complicated. The carrying of heat away from the equator and toward the poles, instead of being accomplished by a simple and steady overturning circulation as in

Figure 2.1.

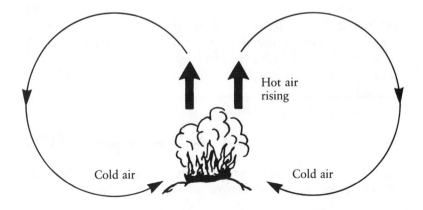

Hot air rising

Cold air   Cold air

figure 2.1, is carried out very unsteadily, with winds frequently changing direction as cyclones and anticyclones develop, move around, and then decay. The westerly winds shown between latitudes 40 and 60 degrees in figure 2.2 look uncomplicated, but they are the average of all the winds produced by such wind systems in the course of a year. They provide only a clue to the wind you are likely to find on a particular day. As we shall see, regions of upward and downward motion accompany the cyclones and anticyclones and the resultant ascent and descent of air produce the main effects of weather. These vertical motions, however, differ from the direct heat-driven buoyant circulations of figure 2.1. Rather, they are forced by the rotating atmosphere's tendency to remain near a balanced state, as intimated in Chapter 1 and as will be explained in Chapter 5.

Over land the world's wind systems are even more complicated because of the very uneven distribution of continents and islands, mountains and valleys, deserts and forests, and other effects large and small that make some areas hotter or colder than others, even in similar latitudes.

Figure 2.2.

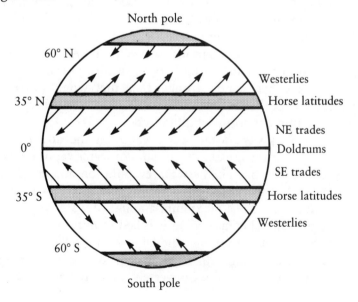

Water vapor

Water vapor is responsible for cloud and fog, suspension in the air of millions of tiny water droplets or ice crystals. It is responsible for much of the active weather we experience, because it moves heat energy around. Water is continually evaporating from oceans, lakes, and moist ground, and this evaporation absorbs heat energy (latent heat). When you hold a wet finger in the wind it feels cold because the process of evaporating the water takes latent heat from your skin. Once the water vapor is in the air it moves around the world, and wherever it condenses into water droplets the same amount of latent heat energy that was used in evaporation is released back into the atmosphere.

The first stage in condensation is the formation of tiny cloud droplets. As condensation continues these both grow and coalesce to make much bigger droplets, which are of course rain. If the temperature is below freezing the result of condensation is likely to be ice crystals rather than water droplets, and the ice crystals grow to form snowflakes. It stands to reason then that areas where rain or snow falls are areas where latent heat is being released, and the heavier the rain the greater the release of latent heat. Tropical cyclones derive much of their energy directly from this release of latent heat. In fact, in middle latitudes as well, this energy source in the rain accompanying cyclones of the belt of prevailing westerlies makes their winds stronger than they would otherwise be. Massive but localized amounts of latent heat energy drive powerful thunderstorms at all latitudes.

There is always some water vapor in the air, and the warmer the air the greater the amount of water vapor likely to be present. The balance between evaporation and condensation depends only on the temperature, and for every value of temperature there is a maximum amount of water vapor that can be present. This maximum is called *saturation*.

If there is saturation and the temperature falls, condensation occurs, either into droplets suspended in the air or onto whatever object is around. Conversely, as the temperature rises there is no longer saturation, and evaporation can commence. Our observation of the typical sequence of damp and dry conditions between night and day helps us understand this. When night approaches, surfaces exposed to the sky cool first and often become damp or even wet as the water vapor nearby condenses. On a grass surface dew forms. The next day, as the sun comes up and the temperature rises, everything dries out again. The temperature at which condensation begins or evaporation stops is called the *dew point*. Put another way, the dew point is the temperature at which saturation occurs.

To the weather forecaster the dew point is an important figure. It is included in every weather report and is essential to the forecasting of both cloud and fog. The criterion for both cloud and fog is that the air should be cooled to its dew point. For instance, if air that is saturated at 70°F is moving over sea whose temperature is only 65°F the air will be cooled below its dew point and fog is very likely.

The difference in the amount of water vapor present at saturation between winter and summer is enormous. At a dew point of 0°F, a not uncommon value in the Midwest in winter, a small room will hold about one ounce of water as vapor. At a dew point of 80°F the same room will hold more than a pound and a half of water as vapor. This explains why tropical storms whose energy comes from latent heat released when water vapor condenses are very sensitive to temperature.

Warming and cooling of air parcels

To understand how all sorts of storms work, we have to see what happens to the temperature of air when it moves up or down, and how the release of latent heat comes into the picture. Sometimes, when thinking about puffy *cumulus* cloud types (see Chapter 6), showers, and thunderstorms, we consider a small sample, or parcel, of air (the size of which does not matter) rising or sinking in a larger environment that is itself nearly at rest. The buoyancy of the air parcel is the key. At other times, when thinking about broad horizontal layers of *stratus* types of cloud (Chapter 6) with gentle but widespread rain, the buoyancy concept is not helpful since the vertical motion is forced on a large scale. That is, the environment itself is displaced upward or downward. It is necessary to know what happens to the layer as it is displaced, and also to know what happens at a fixed height as new environmental air from either above or below replaces what was there initially.

In all these instances we must consider what happens to a parcel of air. When air is let out from a pressurized balloon or tire it cools as it expands. This cooling is a consequence of expansion into the lower pressure of the air outside. When air rises from near the surface it also expands because of the lower atmospheric pressure aloft, and the same kind of cooling occurs. The rate of fall of temperature is called the *lapse rate*.

We can put an exact value on this lapse rate if we assume that the rising air is enclosed in an imaginary envelope that prevents gain or loss of heat from or to the air around. Meteorologists talk of this rising air as a parcel and describe the lapse rate as *adiabatic*. The adiabatic lapse rate of a parcel of air is

Figure 2.3.

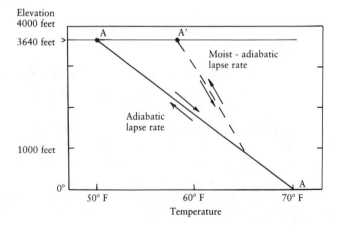

1°F per 182 feet and is illustrated in figure 2.3. As air parcel A rises from the surface to an elevation of 3640 feet, it cools from 70°F to 50°F. If it returns to the surface the opposite happens. The air is compressed, and the compression will cause it to warm back to the higher temperature at the same rate.

There is always some water vapor in the air, but this will not make any significant difference unless saturation occurs. If our parcel of air with a temperature of 70°F has a dew point just below 65°F, the cooling as it rises will bring it to saturation at a height of about 1000 feet. As it rises farther the adiabatic cooling will cause condensation of the vapor and a release of latent heat. This release slows down the cooling rate considerably. Figure 2.3 shows that in this example the air parcel A will arrive at the 3640-foot elevation with a temperature of 58°F, 8°F warmer than it would have been if the latent heating had not occurred.

This slower rate of cooling is called the *saturation* or *moist adiabatic lapse rate*. It is always smaller than the ordinary adiabatic lapse rate, and the difference depends on how much water vapor is in the air at saturation. If the air parcel now sinks it will warm up at the moist adiabatic rate so long as there are enough evaporating water droplets to keep it saturated; the latent heat which was released into the air when the water vapor condensed will be used up again in evaporation. As soon as all the water droplets have evaporated, further descent of the air will result in a warming at the ordinary adiabatic lapse rate.

Stability and instability

In forecasting cumulus clouds, the behavior of a particular parcel, say of the size of an individual cloud, depends critically on the air around it—that is, the environment through which it is rising. Information on this environment is very important to the weather forecaster, and its lapse rate is measured routinely by balloon and satellite. The character of the clouds, the development of showers and thunderstorms, and the gustiness of the wind all depend on this *environmental lapse rate*. Four examples of environmental lapse rate appear in figure 2.4, along with the adiabatic and moist adiabatic lapse rates for a parcel of air rising from point A.

Suppose the lapse rate in the environment is like the one labeled U. The temperature falls with elevation faster than the adiabatic lapse rate A-A for a parcel. Then as parcel A rises from its starting point it immediately becomes warmer and less dense than its environment. It is therefore buoyant and continues to rise at an accelerated rate. In this case the environmental lapse rate is said to be *unstable*, so it is labeled U. The unstable effect is even stronger if the parcel is saturated, so that it follows the moist adiabatic lapse rate A-A'. Vigorous cumulus clouds would develop.

Now suppose instead that the environmental lapse rate resembles the one labeled S in figure 2.4. In this case the rising parcel, even if it is saturated and trying to follow the line A-A', finds itself colder and more dense than its environment and

Figure 2.4. *Various types of environmental lapse rate. A-A and A-A' are the adiabatic and moist-adiabatic lapse rates, respectively.*

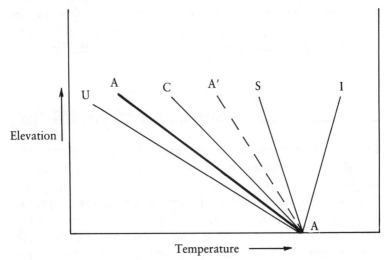

immediately sinks back to where it started. Thus the lapse rate is regarded as *stable*, but less so if the moving air parcel is saturated than if it is not. In either case cumulus clouds would be suppressed. It can be seen that an environment in which the temperature increases with elevation (an *inversion*, denoted by I) is very stable.

When the temperature in the environment decreases at a rate between the ordinary adiabatic and moist adiabatic rates, as illustrated by C, the behavior of the rising parcel will depend on whether or not it is saturated. Only the saturated parcel will be buoyant, and the environmental lapse rate is said to be *conditionally unstable*, again with cumulus clouds the result.

Unstable lapse rates develop over the sea when the temperature of the surface air is much lower than the sea-surface temperature, typically with winds from polar regions. Highly buoyant parcels of air then convey heat and moisture from the sea surface upward to considerable heights, and the air moving downward to take the place of the rising air brings the stronger wind aloft down to the surface in gusts. Typically saturation occurs at a height somewhere between 1500 and 3000 feet, when the rising parcels of air become visible as cumulus clouds. The temperature at the base of the cumulus clouds is the dew point of the air.

Over tropical seas typical environmental lapse rates are conditionally unstable to great heights, even though at sea level the difference between air and water temperatures is slight. Showers are common and thunderstorms occur from time to time. The greatest depth and degree of conditional instability occurs when air from the tropical oceans flows onshore in middle and low latitudes and is exposed to a hot land surface. Violent thunderstorms break out not only over the land but also over coastal waters, such as Chesapeake Bay. Within them the buoyant updrafts may reach speeds of 65 knots.

The continual rising and sinking (overturning) of air parcels modifies an unstable environment so that it becomes less unstable as time goes on, even while addition of heat and moisture at the base of the atmosphere, the surface of sea or land, together with cooling by radiation of the higher layers, is trying to make it more so. On average the environmental lapse rate through much of the atmosphere outside the polar regions tends to be not far removed from the moist adiabatic lapse rate. Much of the cloud we see forms when the environmental rate is less—that is, to the right of the line A-A'—because the broad gentle forced updrafts accompanying the cyclones and anticyclones of mid-latitudes, ranging from a fraction of a knot up to about two knots, occur in more or less stable air. The foregoing buoyancy arguments are of no avail, because the entire environmental layer is lifted and cooled adiabatically until its temperature reaches the dew point. Then a nearly uniform horizontal layer of stratus-type cloud, characteristic of vertical stability, forms over an extensive area. The vigor with which this lifting occurs, however, depends on the degree of stability, as we will see.

Temperature inversions (warmer air over colder) can occur at any height in the atmosphere, but below 35,000 feet or so they are usually shallow, with a depth of not more than about 1000 feet. They represent extremely stable air. Evidence of their presence aloft is sometimes provided by the sudden capping of a cumulus cloud, that is, by the flattening and spreading out of its top just when it looked as though it might develop enough to give a shower. Inversions are common over the sea in summer in middle and high latitudes when the surface air is warmer than the water. In such circumstances the wind at deck level is often surprisingly weak relative to what is measured at the masthead, because mixing with the stronger wind aloft is severely limited. Inversions are also very common over land at night when the ground cools rapidly under a clear sky and cools the air adjacent to it. In such surface inversions over both sea and land, fog is likely to form if the humidity is sufficiently high.

To see how the forced ascent of a layer of air depends on the degree of stability, we must fix our attention on what happens at a particular elevation, say, 10,000 feet, when the air rises or sinks through it as illustrated by figure 2.5. Even though the air aloft is colder than the air below, the environmental lapse rate, indicated by the line S, is stable. The air initially at 10,000 feet is represented by point B.

When the air starting at point B rises it cools adiabatically, as shown in the left-hand part of the diagram. It is replaced at 10,000 feet by the air denoted A, which started from below at a higher temperature. This air has also cooled at the same adiabatic rate on the way up. Since A was warmer than B initially, it is still warmer than B at the final time. At 10,000 feet, however, the indicated cooling, from the *initial* value of B to the *final* value of A, is the *net* of the effect of adiabatic cooling of the rising air itself and the warming effect of initially warmer air rising from below. The ascent of air has produced a cooling at the fixed elevation, but not as strong as the cooling experienced by the moving air itself.

Figure 2.5. *Temperature changes at a fixed height due to the adiabatic ascent or descent of the environmental air.*

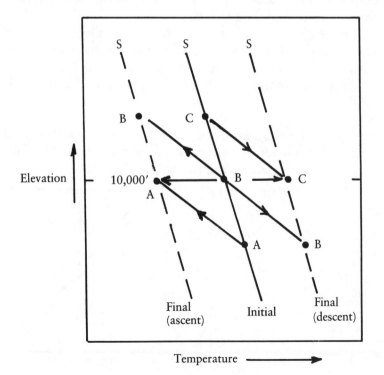

When the air sinks, everything operates in the opposite sense, as shown in the right-hand part of figure 2.5. The air (B) initially at 10,000 feet is replaced by the air initially at point C, which has warmed adiabatically on the way down. The net warming at the fixed elevation is the excess of the adiabatic warming of the moving air over the cooling that would have occurred by simply replacing B with the initially colder C.

The amount of difference between these two effects depends on the stability of the environmental lapse rate, irrespective of whether the air is rising or sinking. Two extreme examples are shown in figure 2.6. Only the ascent is pictured. In the lefthand diagram we show a very stable environmental lapse rate, represented in fact by a temperature inversion. For the same upward displacement of the air as we saw in figure 2.5, a very large cooling results at 10,000 feet, because the air initially at A, besides cooling adiabatically as it rises, was colder than B to begin with. The cooling at 10,000 feet is now the *sum* of two cooling effects.

In the right-hand diagram we see what happens when the environmental lapse rate is itself nearly the adiabatic value. This time very little cooling results at 10,000 feet for the same upward displacement of air. The final value of the air (A) is

Figure 2.6. *Temperature changes for the same adiabatic ascent of air when the environmental lapse rate is very stable* (left) *and only slightly stable* (right).

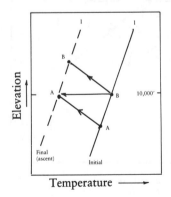
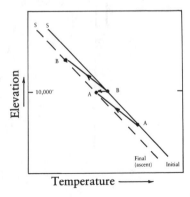

only slightly colder than the initial value (B) at this elevation. This happens because A was so much warmer than B initially. The net cooling is the small excess of the adiabatic cooling of A over the initial relative warmth of A compared with B.

Let's look at the same thing from a slightly different point of view. Suppose that the response to large-scale forcing of the type to be described in Chapter 5, in order to restore near-balance, requires a given amount of cooling at 10,000 feet. Suppose it is the amount shown in figure 2.5 but suppose further that the environmental lapse rate is extremely stable, as shown in the left-hand part of figure 2.7. Not much vertical displacement would be required. That is, the response to a given forcing is minimal, and it is fair to say that the atmospheric "spring" is "stiff" when the stability is large. On the other hand, when the environmental lapse rate is close to the adiabatic value, as in the right-hand portion of figure 2.7, a large vertical displacement is required to produce the needed cooling. In this case the response to a given forcing is generous. The atmosphere is "limber" when the stability is small.

We are now prepared to see what happens when the air is saturated with water vapor and ascent results in the release of

Figure 2.7. *Amount of adiabatic ascent required to produce the same environmental cooling in very stable air* (left) *and in only slightly stable air* (right).

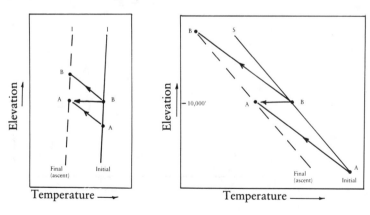

latent heat of condensation. The situation is pictured in figure 2.8. The environmental lapse rate is the same as in figure 2.5, but now the rising air cools at the moist adiabatic lapse rate. The same vertical displacement of air as in figure 2.5 now results in much less cooling at 10,000 feet. The reason is that the latent heating partially offsets the ordinary adiabatic cooling. The effect is the same as if the air were unsaturated but less stable. The release of latent heat always makes the atmosphere more "limber" than it would otherwise be. For a given forcing the updraft is more vigorous and so is the surface cyclone, as

Figure 2.8. *Temperature change produced when the ascent and environmental lapse rate are the same as in figure 2.5, but the air is saturated.*

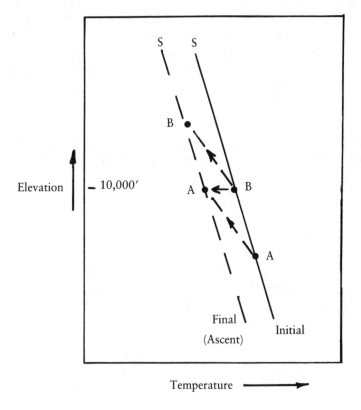

we will see in Chapter 5. This is a reason why cyclones with massive amounts of clouds and rain in air with only slight vertical stability, as is typical over the oceans in winter, are likely to be much more vigorous than cyclones with sparse available moisture in strongly stable air.

We can perhaps visualize what would happen if the environmental lapse rate were exactly the same as the adiabatic lapse rate, either ordinary as in the right-hand side of figure 2.6 or moist as in figure 2.8. No amount of vertical displacement would change the temperature at 10,000 feet at all, because the adiabatic cooling of the air coming up from below would be exactly balanced by its excessive warmth initially. If the environmental lapse rate were actually unstable (see figure 2.4) the ascending air would be warmer than the initial air at 10,000 feet. In this case there would be no orderly gentle ascent of air on the large scale but rather the buoyant ascent and descent of individual small parcels, as we discussed earlier, resulting in cumulus clouds if the air becomes saturated.

In cyclones, cumulus convection is often superimposed on the more orderly forced updrafts and downdrafts without disrupting the main cyclonic circulation. This is because the air is unstable only in limited sectors of the low-pressure area or in layers of limited vertical extent. The uniform layered appearance of the sky suggests forced lifting of stable air, but bursts of torrential rain and perhaps thunder in the midst of a steady lighter fall are a sign that regions of instability and buoyant cumulus clouds are embedded in the major mass.

3

Weather Basics— Mapping the Winds

This chapter shows how sea-level isobars provide clues to the strength and direction of the wind; why, more often than not, we can't map the winds precisely from isobars alone; and why the wind far aloft may differ markedly from the wind near the sea surface. Like Chapter 2, this chapter builds toward Chapters 4 and 5, and an understanding of how weather systems form.

The easiest way to map the winds of the world is first to map the pressure (or weight) of the atmosphere at sea level. Satellite pictures complement this method, particularly over parts of the oceans where surface observations are few and far between. For instance, they often show the pattern of lines of low cloud oriented along the wind direction, and given suitable targets and computer processing, individual clouds can be tracked and estimates obtained of the direction and speed of the wind in which they are embedded.

Much can be deduced, however, simply from a pattern of *isobars* (lines through points with identical values of pressure, analogous to contours of elevation on a topographic map). A simple example is shown in figure 3.1. Where friction does not have to be taken into account, there is a direct relationship between the orientation and spacing of the isobars and the wind. If the wind is steady it is nearly the same as what meteorologists call the *geostrophic wind*, a hypothetical wind which would blow along the isobars if there was an *exact* balance between the two forces acting on the air. These are the pressure-gradient force, directed toward lower pressure, and the

Coriolis force, acting to the right of the wind flow in the northern hemisphere. The term "geostrophic" means "earth-turning," a reference to the Coriolis force, which makes the wind parallel the isobars rather than blow perpendicular to them. You can see from figure 3.1 that the geostrophic wind must blow exactly along the isobars. There is an easily remembered rule—Buys-Ballot's Law—that tells you which way it blows in relation to the pressure gradient. It says that if you stand with your back to the wind, you will have low pressure on your left-hand side (figure 3.2). In the southern hemisphere the Coriolis force is reversed, and so of course are the wind direction and Buys-Ballot's Law.

The Coriolis force at a given latitude is directly proportional to the wind speed, so closely packed isobars on a weather map mean that a large Coriolis force is needed to balance the powerful pressure-gradient force (figure 3.1) and

Figure 3.1. *Relationship between horizontal pressure gradient force, Coriolis force, and geostrophic wind.*

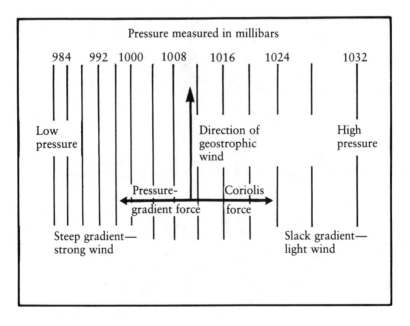

the wind is strong. On the other hand, widely spaced isobars indicate light winds. The Coriolis force also depends on the latitude. For a given spacing of isobars it is a maximum at the pole and disappears at the equator. So the same spacing of isobars gives more wind the farther south you go. For example, if isobars drawn at intervals of four millibars (as in figure 3.1) are 200 nautical miles apart at latitude 45° N, the geostrophic wind is 17 knots. The same spacing at 25° N gives 28 knots. If you have developed a feel for wind and isobaric spacing in one latitude, you may be surprised by conditions when sailing in another.

The geostrophic wind may seem like an abstraction, but the direction and speed of the actual wind above the lowest 3000 feet or so is usually close to it and a good fit to the isobars on the sea-level weather map. If low clouds are visible, always make a habit of observing how they are moving when you are on land or in harbor, because they will usually give you a good idea of the wind offshore.

Figure 3.2. *Buys-Ballot's Law.*

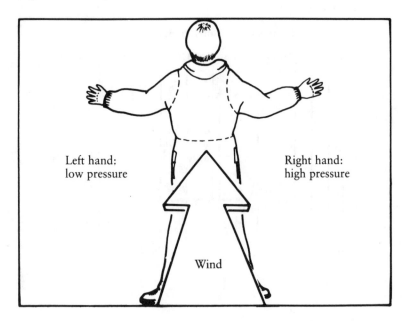

Left hand:
low pressure

Right hand:
high pressure

Wind

The wind near the earth's surface is slowed down by friction, and its direction is altered so that it blows at an angle across the isobars toward lower pressure (figure 3.3). Over land the surface speed (actually measured at an elevation of about 30 feet) may be less than half the geostrophic value, and the turning angle may exceed 45 degrees. Over the open sea, which is smoother than land even when the seas are rough, the angle is about 15 degrees and the average speed is about 75 percent of the value measured from the isobars.

The relative temperatures of air and water can make a big difference. In an unstable situation when cold air is being strongly heated by the sea, the mixing in the air is vigorous and the friction effect is spread through a large depth of the atmosphere. Then the direction and speed of the surface wind can be close to the geostrophic values. When warm air overlies cold water and a stable situation results, mixing is inhibited, the friction effect is concentrated in a shallow layer, and the sur-

Figure 3.3.

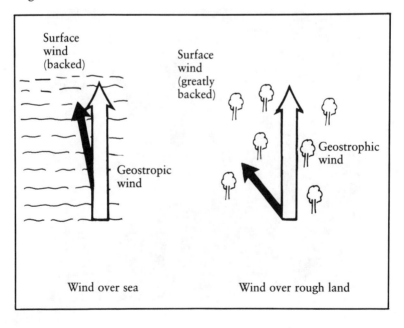

Surface wind (backed)

Surface wind (greatly backed)

Geostropic wind

Geostrophic wind

Wind over sea   Wind over rough land

face wind is weaker and turned more toward lower pressure. In the strong temperature inversions that are often found over cold waters during the sailing season in middle latitudes, wind on the sea surface can be almost calm despite moderate speeds at the level of the masthead anemometer.

In coastal waters a change of water temperature between 5° and 10°F in a few miles is not uncommon, and the resulting change in wind speed and direction can be significant—as much as 10 degrees in direction and 25 percent in speed. Across the inshore edge of the Gulf Stream, the change between stable and unstable conditions can be abrupt. With the same spacing of isobars a strong wind and rough sea in the Stream

Satellite picture showing many lines of cloud. They are oriented mostly along the direction of the surface wind or the wind aloft. A solid layer of summer fog and low cloud outlines the coast of California and intrudes into some of the bays and valleys.

can give way to nearly calm conditions over a distance of a mile or less into the colder waters of the continental slope.

Over land, and over rivers as well as over all but the largest lakes, there is a 24-hour cycle in the wind strength caused by a cycle in the stability of the boundary layer. The sun heats the ground and thus the air near it, in the way that a warm sea heats cold air over it. The surface wind increases because the resulting mixing brings down some of the stronger wind higher up. As the sun goes down, the temperature falls and the wind decreases. During the night, if there is little or no cloud, the ground cools rapidly, the air becomes very stable, and the wind quickly dies away.

Over the sea there is little of this diurnal cycle, since the surface water temperature changes less than a degree over the course of the day. An average of 56 days of sailboat masthead anemometer readings by one of us during spring and summer passages in the western Atlantic (figure 3.4) shows only about one knot of increase from an early morning low to a late afternoon or early evening high. The likelihood of light air, less than 10 knots, however, was 40 percent greater from 0400 to 0800 (ship's time) than from 1600 to 2000, while strong breezes of 25 knots or more were 30 percent more frequent from 2000 to 2400 than from 0400 to 0800.

Wind variability over the course of a day at sea masks these small diurnal effects, as illustrated by two examples in figure 3.4, in which the anemometer was read at 15-minute intervals. During the passage to Bermuda, generally light air over the cooler slope water built to a fresh breeeze of 25 knots or so in a series of bursts and squalls crossing the Gulf Stream into the Sargasso Sea, with showers and signs of thunderstorms. The more sedate passage to Halifax showed a general decrease of wind, with mist and fog during the early morning hours, despite winds of 10 to 20 knots, clearing during the day. Then during the evening the wind collapsed and fog set in on the approach to Nova Scotia. In both cases the wind varied on all time scales down to the basic 15-minute observation interval.

The sea-level isobars in these cases gave information about the direction and general strength of the winds, except when they were extremely light, and showed only the trend over the day as a whole. In general, even after due allowance is made for friction, the winds will differ by as much as a few knots from an estimate based on the isobars. This is because of the presence of vertical circulations, like the one shown schematically in figure 2.1. The winds as given by the isobars have no significant convergence or divergence and so when vertical circulations are present there must also be a weak *nongeostrophic* wind flow across the isobars to provide continuity. Except in the most unusual situations, the main part of the wind is the part that can be inferred from the isobars. Thus, while the isobars are very useful, they are not the whole story, especially in matters of detail.

Figure 3.4. *Winds speeds from a masthead anemometer: hourly averages compiled from readings taken over 56 days (heavy line) compared with individual readings on two sample days. There is, on average, little diurnal variability over the sea, and what little there may be is easily masked by nondiurnal variability.*

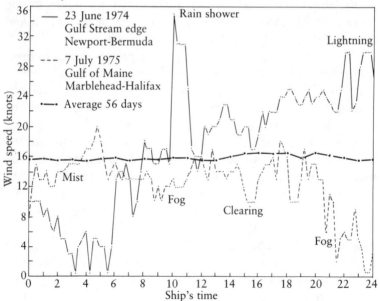

_____ Wind shear up the mast

Because of the effect of surface friction, as we saw before, the nearer you are to the surface of land or sea the slower the wind and the more its direction is angled across the isobars toward lower pressure. The change in speed is noticeable even between the top of the mast and deck level. Even though the direction of the *true* wind changes only slightly in the lowest hundred feet, the direction of the *relative* wind felt by you and by the sails can change significantly between the deck and the masthead. Differences in speed range from very little in unstable air (about five percent) to enormous amounts in stable air (up to 300 percent difference).

Let's examine the difference in relative wind direction when you are sailing at 5 knots with a true wind on the beam at a speed of 15 knots, as measured at the masthead. Then a little trigonometry will show you that the relative wind direction at the masthead, as measured by an apparent wind indicator, will be 18 degrees ahead of the beam. In unstable air you might have at deck level a true wind still on the beam at 14 knots, giving a relative wind 20 degrees ahead of the beam, a difference of only 2 degrees. In stable air, on the other hand, the true wind might still be on the beam, but at a speed of only 5 knots. In that case the relative wind would be 45 degrees ahead of the beam, a 27-degree difference in the relative direction between deck and masthead! So in stable air you will need a large amount of twist in your sails, while in unstable air the leeches can be much straighter.

Other examples can give surprising results. Our favorite is an experience sailing downwind over the cold waters of the Maine coast. We were going 6 knots with the spinnaker set, and 6 knots was showing on the masthead anemometer. The foot of the spinnaker was fluttering, however, because a slight *head* wind was blowing under it at deck level.

This wind shear largely accounts for what is known as "weight of wind." Variations in wind speeds between the bottom and top of the mast can give something like a 50 percent difference in the heeling moment on the boat for the same wind speed over the deck. In stable air there will appear to be more "weight in the wind" for a given wind speed over the deck, or less if the given wind speed is measured by the masthead anemometer. In stable air, moreover, the sea will be relatively smooth for a given masthead wind speed. In unstable air, as in outbreaks of cold air up north or at any time over the warm water of the tropics, the roughness of the sea for a given masthead speed will come as a surprise for someone accustomed to summer sailing in cold water.

_____ Wind shear higher up— thermal wind and the jet stream

Important as the practical considerations just discussed may be, our understanding of how weather systems work depends on considering the vertical change in the wind in the depth of the atmosphere above the surface friction layer. It must be clear by now that it is impossible to understand or even fully appreciate weather merely by studying the pattern of surface winds or the standard weather map with its isobars. Events at the surface are linked with events throughout the depth of the atmosphere, and the winds we use for sailing are an integral part of a 3-dimensional pattern of air movements driven by a very complicated heat engine. We could give up as we begin to realize the complexity of it all, and decide merely to sail with the wind as we find it, with the wind that "bloweth where it listeth." But this would remove one of the fascinating dimensions of sailing. The winds *can* be understood, and, as we shall see in Chapter 6, even the clouds may tell their own story of events large and small in the atmosphere around us.

We have looked at how isobars represent the surface wind pattern. We have taken a satellite-eye view of weather systems, and we have seen how on the small scale the temperature of the air determines whether it is likely to erupt in unstable convective updrafts and downdrafts. To understand the consequences of the change of wind as we go up through the atmosphere above us, we will need one more concept, that of the *thermal wind,* which produces wind shear effects at altitudes higher than those affected by surface friction.

To find out about the thermal wind, we start by realizing that within the same vertical distance cold air, which is relatively dense, must be heavier than warm air, which is less dense. This implies that pressure decreases more rapidly with altitude in cold air than in warm air, for the simple reason that pressure at a given level measures the weight of the air above. By the same reasoning, pressure increases more rapidly with decreasing altitude in cold air than in warm. So if we know the average temperature of the air in the vertical and how it varies from place to place, we can infer how the pressure pattern changes with height. What does this mean so far as the wind is concerned?

Consider the simplified possibilities illustrated in figure 3.5, in which we ignore the effects of friction for the purpose of argument. Suppose first that the pressure is horizontally uniform aloft and the air is not moving. Then the weight of the colder air will produce relatively high pressure at the surface compared with conditions where the air is warm. If the air is not moving at the surface, it will accelerate directly across the isobars from high to low pressure. Immediately the Coriolis force will cause the resulting north wind to turn to the right, and in the final state of balance the air will be moving as an east wind along the isobars in geostrophic balance, with the high pressure (and the cold air) to the right. If conditions have remained uniform and still aloft, then the vertical *shear* of the wind (that is, the calm aloft *minus* the final geostrophic wind at the surface) will be parallel to the isotherms with the cold air to the left. This vertical difference is the thermal wind. It is westerly in this case.

Now let's see what happens when the pressure at the surface is uniform and the wind is calm. Then the greater weight of the cold air will lead to low pressure aloft while pressure is relatively high in the warm air. The initially still air aloft will accelerate northward, turn to the right under the influence of the Coriolis force, and end up as a geostrophic west wind, with low pressure and cold air to the left. The thermal wind is now the difference between the westerly aloft and the calm at the surface. As in the preceding example, the thermal wind is from the west, with the cold air on the left.

In the real atmosphere, the situation is more complicated. Some combination of the two examples shown above is nearer to reality, but the balanced geostrophic wind can be blowing with a component across the isotherms as well as along them. The range of possibilities is illustrated in figure 3.6. When, as shown by the isotherms, colder air lies to the left of the lower-level wind, the thermal wind is in the same direction and the upper-level wind is stronger. If colder air is to the right, then the low-level wind weakens with elevation and may reverse direction because the thermal wind opposes it. If the low-level

Figure 3.5. *How the thermal wind would develop, from a state of rest to a balanced equilibrium.*

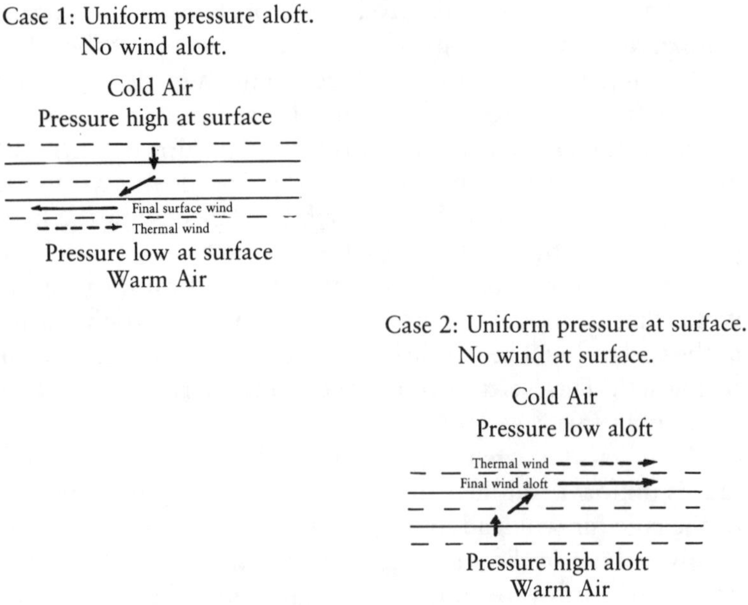

Case 1: Uniform pressure aloft.
     No wind aloft.

          Cold Air
     Pressure high at surface

          Final surface wind
          Thermal wind
     Pressure low at surface
          Warm Air

          Case 2: Uniform pressure at surface.
               No wind at surface.

                    Cold Air
               Pressure low aloft

          Thermal wind
          Final wind aloft

               Pressure high aloft
                    Warm Air

wind blows from warmer air toward colder (*warm air advection*) then the wind turns clockwise (veers) at higher levels and strengthens, because the thermal wind is to the right. When the lower wind blows from colder toward warmer air (*cold air advection*) the picture is the same, except that the direction turns counterclockwise (backs) with elevation, because the thermal wind is to the left.

Notice that the thermal wind always blows along the isotherms, just as the geostrophic wind blows along the isobars. Like the geostrophic wind, it is stronger the more closely spaced are the isotherms and the lower the latitude. There is even a thermal Buys-Ballot's Law: if you stand in the northern hemisphere with the thermal wind at your back, cold air is on your left-hand side.

Figure 3.6. *Some idealized examples of thermal wind.*

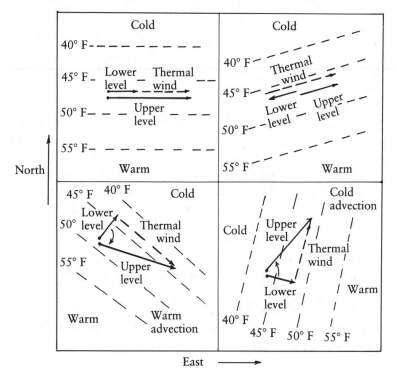

The scenarios depicted in figure 3.5 are unrealistic in that they presume that the air starts from rest, far from geostrophic balance. Actually the air flow in cyclones, anticyclones, and fronts remains fairly close to geostrophic balance all the time. But when the temperature contrasts change, cross-isobaric flow occurs, to keep the balance close. For example, when the isotherms come closer together, as they do when a front is forming, then the response is qualitatively like the picture shown in the figure. That is, the surface flow crosses the isotherms toward the warm air while aloft the air moves toward colder air. At the same time, as we will see in Chapter 5, vertical motions produce adiabatic cooling in the warmer air and warming in the colder air. Thus the increased temperature contrast eases in part as the wind accelerates toward the new geostrophic balance.

Jet streams, well known to aviators for their influence on the speed of crossing the Atlantic and Pacific oceans, are a consequence of the thermal wind. They are found where deep layers of strongly contrasting warm and cold air lie adjacent. Normally, colder air lies to the north, so jet streams blow from the west, but in a given weather situation the temperature pattern and the jet stream can be distorted into a variety of directions and shapes.

As with the geostrophic wind, the thermal wind is an abstraction, but it can be sensed from the shore, from a slowly moving boat, or from a satellite film loop when two or more layers of cloud are present. Try observing the relative motions of a layer of low clouds and a layer of high ones. Condensation trails from jet aircraft fulfill the same role as high clouds in revealing the wind aloft. You will sometimes observe a large shear—that is, a strong thermal wind—and it will be useful to interpret it in terms of warm or cold air advection using the models in figure 3.5. For reasons to be explained in Chapter 5, a strong thermal wind to the right of the surface wind (warm advection) often heralds bad weather, while a strong thermal wind to the left of the surface wind (cold advection) will indicate clearing weather.

4

The Structure of Weather-Map Systems

This chapter examines sea-level weather maps and their counterparts aloft and offers a glimpse of how temperature, pressure, and wind interact between the two levels.

We are now in a position to look at weather maps, both at the surface and aloft, in terms of either pressure or wind, and to appreciate the temperature patterns that link the flow patterns at the two levels. An example of many of the things we have been talking about is presented in figures 4.1–4.3.

At the surface, visualize in figure 4.1 the isobars of equal sea-level pressure that accompany the whirls and broad currents of wind with Buys-Ballot's Law in mind. Thinking of them as though they were contours of land elevation, we find a gentle mountain of high pressure at the center of each outward rotating whirl, or anticyclone. At the center of each inward rotating cyclone lies a hollow of low pressure, sometimes with very steep sides. Gales are blowing where the isobars are most closely packed.

Where the isobars bend sharply around these features, we find ridges if the flow turns clockwise and troughs if the turn is counterclockwise. Elongated troughs of low pressure are seen connecting some of the neighboring surface lows and separating highs in the checkerboard pattern of centers. Warm and cold masses of air impinge along these trough lines, and we have the particular form of trough known as a frontal trough or front.

At upper levels (figure 4.2), 18,000 feet elevation, the "high land" is in the tropics to the south, and the Arctic basin is a great broad hollow of low pressure with numerous individual low centers, some of which extend into middle latitudes. The broad slope of pressure, across which the westerlies blow, following the isobars, is indented by roughly north-south valleylike upper troughs and intervening upper ridges, corresponding to the clockwise and counterclockwise bends in the current. Where the slope is steep, the upper current is concentrated into a swift jet stream.

Figure 4.1. *Surface map. The thin lines are isobars of pressure at sea level, and the heavy lines are streamlines of surface wind flow.*

The *difference* between the patterns of geostrophic wind, as expressed by the isobars, at the two levels tells us the mean temperature of the layer from the surface to 18,000 feet (figure 4.3), and also the thermal wind. Recall Buys-Ballot's Law for the thermal wind and isotherms of temperature. The general tendency for westerly flow aloft is due to generally warmer air in low latitudes than farther north. Comparison of figures 4.3 and 4.2 shows that the temperature contrast is especially intense across the localized regions of the jet stream. Comparing figures 4.3 and 4.1 we see that bands of strong temperature gradient lie parallel to the surface frontal troughs and poleward from them. Thus, zones of strong horizontal temperature

Figure 4.2. *Isobars at an 18,000-foot elevation. The wind almost exactly parallels them.*

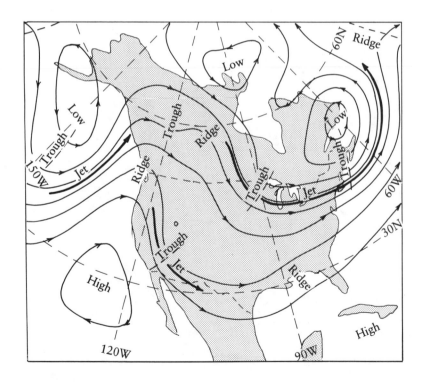

gradient (provided they are not too shallow), surface fronts, and upper-level jet streams are found together in this typical arrangement.

By a careful comparison of figures 4.1 and 4.2 you can see that the various features of the sea-level isobaric pattern have their counterparts aloft, typically displaced to the west about 10 degrees of longitude. You can think of this displacement as being linked to the distortion of the temperature isotherms produced by the circulation of wind around the cyclone or anticyclone. For example, the southward plunge of cold air in the northerly flow west of a surface cyclone produces a cold trough in the temperature pattern. This is reflected in an upper

Figure 4.3. *Isotherms of temperature averaged over the layer from the surface to 18,000 feet. The cold air aloft makes the average cold.*

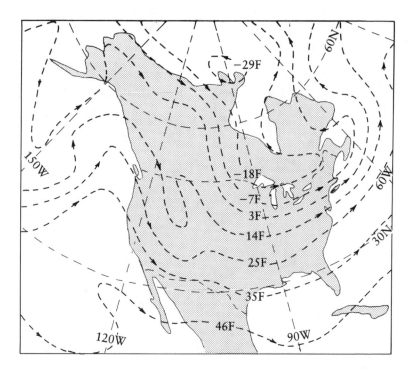

trough in the flow aloft west of the surface center. Conversely, the circulation of air around a high tends to produce warmer air to the west of the surface center than to the east, with a corresponding westward shift of the accompanying upper ridge. We will see in Chapter 5 that the presence of an upper trough west of a surface cyclone is crucial for its development.

Usually the centers of high and low pressure at the surface are reflected by ridges and troughs aloft rather than by upper centers. The young low near 100° W is accompanied by an upper trough at about 110° W, and the two highs in North America as well as the low in western Canada are each accompanied by a westward-displaced upper ridge or trough. On the other hand, the deep surface lows near 60° W and 140° W have counterpart lows at the upper level at nearly the same location, with jet streams circling the periphery. This structure is characteristic of old mature cyclones, which are not likely to intensify further but may persist over the oceans for some days.

When we look at a sequence of maps like these, with the accompanying satellite cloud pictures, we see the evolutions discussed briefly in Chapter 1 and illustrated in figures 1.1–1.3. Next we are going to explain the mechanisms that underlie the behavior of cyclones, anticyclones, upper troughs and ridges, and fronts.

5

Weather-Map Systems— How They Work

Chapters 2, 3, and 4 are pulled together here in an explanation of how low- and high-pressure systems form, how they move, and why typical weather patterns are associated with each. We will also see how warm and cold fronts are formed.

We now have an idea of how cyclones (lows) and anticyclones (highs) are always found on the weather map in considerable numbers. Some last only a day or two and never advance beyond a feeble infancy; some, especially the larger systems, may last for a week or more. Some move quickly—they are the young systems beneath the the fast-flowing jet streams in mid-latitudes. Some remain nearly stationary—particularly the larger cyclones over the subpolar oceans and anticyclones over the continents in winter or the subtropical oceans throughout the year. The Azores and Eastern Pacific highs are examples of the latter, and are of particular interest to sailors as they slowly drift along and weaken or strengthen. There is an infinite variety of cyclones and anticyclones. No two weather systems are ever identical, so in describing them we must be satisfied with describing typical features and behavior.

It is something like watching the flow of a stream. There is the main current, normally but not always near the center, while eddies develop and are carried downstream as they slowly weaken. Many eddies form where rocks interrupt the flow, just as in the earth's atmosphere many cyclonic eddies

44

form to the lee of major mountain ranges, preferably in a jet stream. These orographic eddies persist as prominent upper troughs as they move downstream. The corresponding orographic cyclone at the surface often remains weak and is trapped near the mountains, moving southward along the lee slopes. The major surface cyclones develop when one of these upper troughs approaches a separate zone of strong temperature contrast in the near-surface layers of the atmosphere. Such a zone typically lies in a nearly stationary frontal trough, a favorite location for which is along or near a continental eastern coastline. In winter the land is colder than the sea and even in summer there is an appreciable contrast in the western Atlantic between the cool air over the water of the continental shelf and the warm air over the Gulf Stream and the Sargasso Sea. The surface cyclone forms in this zone of strong temperature contrast through a progressive linkage with the upper trough, as we shall see later in this chapter.

In a stream the energy for the eddies comes from the energy of the main flow as the water runs downhill. That is, the center of gravity of the water becomes lower downstream, and the released energy is used to maintain the flow against friction of the bottom and the banks.

In the atmosphere the situation is different. The energy for the winds of an eddy comes from an average sinking of colder, denser air and ascent of warmer, lighter air in the neighborhood of the weather system. This tends to lower the center of gravity of the total mass of air. The energy provided by this lowering drives the winds of the eddy and maintains them against surface friction. The additional energy produced by the release of latent heat of water vapor in the great masses of cloud and rain, and the energy provided to the atmosphere by the transfer of heat and moisture from a warm sea surface, can enhance the vigor of the process and lead to stronger winds. But the process of mid-latitude cyclone development itself (as distinct from tropical cyclone formation) requires *horizontal contrasts* of warm and cold air at the same elevation. In fact, the examples of cyclones, anticyclones, fronts, and upper troughs and ridges that we will present all depend on horizontal temperature gradients for their typical behavior, and indeed for their existence.

At the same time that these processes are going on, part of the energy of the winds in the developing eddy are being returned to maintain the strength of the overall westerlies aloft in middle latitudes. And the sun continues to heat the low latitudes more than the northern regions, so as to maintain the atmospheric flow and the accompanying temperature contrast which makes possible the development of more cyclones.

Let us see in more specific and practical terms how the processes of ascent and sinking of air work in a cyclone and how the circulation develops and moves. The central idea is

Cyclonic cloud vortex development over the North Atlantic. A mature vortex can be seen far at sea, while a newer one is growing off the east coast of North America.

that the atmosphere, although close to balance, is continuously being nudged out of geostrophic and thermal-wind equilibrium by its own advections of temperature and wind momentum. The atmosphere is just as continuously acting to reestablish that equilibrium.

To restore this balance something must happen which cannot conform, naturally enough, to our simple notion of a pure geostrophic and thermal wind. There has to be an additional movement of air, a wind, usually across the isobars. We showed an example of this cross-isobaric wind and acceleration in the establishment of the thermal wind in Chapter 3. This additional component is called the nongeostrophic wind, and it is linked to the broad and gentle updrafts and downdrafts whose existence we inferred from the satellite pictures in Chapter 1. The linkage works qualitatively in the same way as the inflow and outflow accompanying the violent fire-driven updraft in figure 2.1. The adiabatic warming and cooling brought about by these modest vertical currents, together with the wind accelerations due to the nongeostrophic wind components, are what restore the balance.

The strength of this nongeostrophic circulation is directly proportional to the extent to which the balance has been upset and to the strength of the temperature gradient. Furthermore, it is inversely proportional to the vertical stability of the atmosphere. Actually, the upsetting and restoration of the balance is going on more or less simultaneously, so that the atmosphere in general stays near geostrophic equilibrium all the time. Although the nongeostrophic wind component rarely reaches more than a few knots and although the vertical drafts rarely reach even one knot, this circulation is crucial in determining the pattern of weather associated with the typical sequence of lows and highs experienced in temperate latitudes.

A single general rule is based on this central idea. This rule asks us to look at how the geostrophic wind we would measure, as shown by the isobars, changes when we proceed *along* the isotherms with the thermal wind at our back (that is, with the cold air to our left according to the thermal Buys-Ballot's Law). Let's see how it works out in a variety of typical weather patterns.

Lows and highs

Suppose we look at a simple case of straight east-west isotherms on a section of the weather map. In the vicinity of a cyclone, the change along the isotherms will be from a northerly wind to a southerly wind as we walk, so to speak, across the low center (figure 5.1). Cold air moving southward where there are northerly winds will produce a cold trough, and warm air moving northward where there are southerly winds will produce a warm ridge. This distortion in the isotherms, however, means a distortion in the pattern of thermal wind, upsetting the geostrophic balance we started with.

The low-level nongeostrophic component that comes about to restore the balance flows at right angles clockwise to the geostrophic change vector as shown in figure 5.1. In this case it blows from west to east across the low center. This situation is similar to the example shown in figure 3.5. The thermal wind in figure 5.1 *changes* from the earlier state of westerly to a later state of southwesterly—that is, the change

Figure 5.1. *How to deduce the low-level nongeostrophic wind when a low is embedded in a horizontal temperature gradient. The solid oval is an isobar and the dashed lines are isotherms, heavy representing an early state and light a later one.*

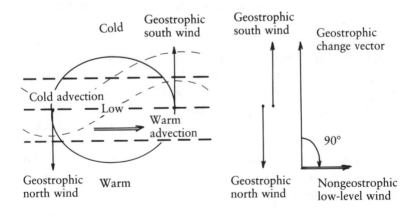

is in the southerly sense. Consistent with figure 3.5, the low-level nongeostrophic wind is westerly, across the thermal-wind change from cooling toward warming.

In highs the picture is a mirror image of figure 5.1, and putting the two together in figure 5.2 we see that the non-geostrophic winds converge near the surface in the southerlies carrying warm air northward east of the low and west of the high (warm advection). They diverge in the northerlies carrying cold air southward to the west of the low (cold advection). Convergence in the surface winds implies ascent—the air can go only upward—and ascending air means condensation, clouds and rain. Divergence in the surface winds on the other hand implies subsidence—that is, air drawn down from aloft—and subsiding air means warming leading to evaporation, drying of the air and clearing skies.

Near the earth's surface there is a second cause of air moving across the isobars, which we have already mentioned in earlier chapters, namely, friction. The consequence of friction upsetting the balance between the pressure gradient and the Coriolis force is a flow across the isobars toward lower pressure. This produces a very weak general ascent over the area of a low and descent over a high.

Putting the two patterns of ascending and descending air together we arrive at an understanding of the weather typically observed in lows and highs: major cloudiness and rain in the region of warm advection to the east of the low, extending

Figure 5.2. *Inferring vertical motion and weather in a sequence of cyclones and anticyclones. The solid and dashed lines show a typical isotherm and isobars, and the heavy arrows represent low-level nongeostrophic wind.*

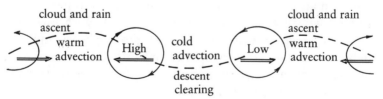

westward over the area of the center, and clearing skies in the region of cold advection to the east of a high (west of a low), with fair weather persisting over the high center itself.

To understand how lows move we must add one more idea: The low is a mass of cyclonically rotating air, spinning counterclockwise around the center. Air converging toward a central point is deflected to the right by the Coriolis force. That is, convergence of the air causes the spinning to speed up, in the same way that a spinning ice skater rotates faster by drawing arms and legs in as close as possible to the center of the body. In the case of both the cyclone and the ice skater, divergence produces a slowdown of the rotation. The low moves from the region where cyclonic rotation of air is slowing down toward the region where it is speeding up. In figure 5.1 this is from left to right, from where cold air is advancing (cold advection) toward where there is warm advection. Since the cyclone is also a region of low pressure, pressure is rising where there is cold advection (with clearing skies) and falling where there is warm advection (with cloud and rain).

You may have noticed that this explanation accounts for the traditional rules about barometric pressure and weather: fair with rising pressure and foul when the barometer is falling. The friction effect accounts for low pressure itself connoting bad weather and high pressure fair conditions, but the main relationship is not so much between weather and pressure as between weather and pressure *change*.

This whole picture has been shown for the situation where the isotherms run west-to-east. It is equally valid when the isotherms are oriented in any direction at all. Lows (and highs) move along the thermal wind regardless of its direction, with cold air to the left of track and warm air to the right. It is as though they were steered by the thermal wind—or to a close approximation, by the wind aloft, which tends to parallel the thermal wind (compare figures 4.2 and 4.3). With lows, the isotherms are *usually* running southwest to northeast, and near highs they *usually* are oriented from northwest to southeast. Hence, on the average but not always, lows move toward the northeast and highs toward the southeast. All the pressure centers in figure 4.1 follow this guideline except the weak lows near 30° N and 60° N. The isotherms in these cases run northwest-southeast (figure 4.3), and the centers are moving southeastward.

Upper-level troughs

Often the isotherms are not approximately straight, but are distorted into a series of prominent *cold troughs* and *warm ridges*. At upper levels (roughly between 10,000 and 30,000 feet of altitude) the temperature advection together with adiabatic warming and cooling tend to produce structures in which the isobars conform closely to the isotherms, and so we see pronounced cold troughs and warm ridges in the flow pattern. Examples can be found in figures 4.2 and 4.3. An idealized situation of this type is shown in figure 5.3, in the vicinity of a trough. As this structure progresses eastward the air must cool in its path and warm behind it. Since there is no temperature advection to produce the needed change, adiabatic cooling and

Figure 5.3. *How to deduce the low-level nongeostrophic wind when there is an upper-level cold trough. The solid line is an isobar and the dashed lines are isotherms, heavy representing the initial state and light a later one.*

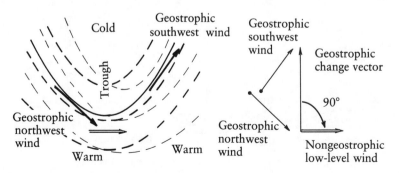

warming must do the job. There must be ascent ahead of the trough and subsidence to its rear. Although the isotherms are not straight, the same principle that applies to figure 5.1 applies here also, and (figure 5.4) is the reason why clouds and rain often precede the passage of an upper-level trough while clearing follows it.

If the upper trough does not extend down to the surface, then the clouds are high and broken and the rain not of any consequence. If the trough does extend to the surface with some strength, as it often does in the rear quadrants of mature intense cyclones (see the examples in figure 4.1), then strong squally winds and brief heavy downpours may precede and accompany its passage.

Cyclogenesis— the birth of a low

The upper-level trough has a further significance, however. If there is a strong temperature gradient near the surface in the region of convergence ahead of the trough, then a new cyclone is likely to form as surface winds begin to spin at an increasing

Figure 5.4. *Inferring vertical motion and weather in a sequence of upper troughs and ridges. The solid and dashed lines are isobars and isotherms.*

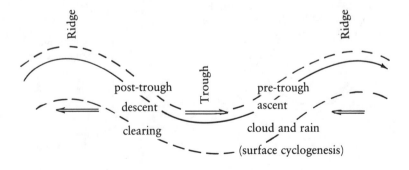

rate about the developing center (figure 5.4). *Cyclogenesis* has then occurred.

The circulation near the surface will continue to intensify as long as the ascent aloft over the center due to the upper trough to the west provides enough convergence, through the nongeostrophic wind, to overcome the surface frictional drag on the rotating winds. After a day or so, the upper trough overtakes the surface storm, and itself develops a complete cyclonic circulation as it is fed in return by the now vigorous cyclone below. At this stage the combined system has reached maturity and moves together for a while (slowly northeastward as a rule) in a steady state before beginning a slow period of decay that may last for a number of days. The deep combined lows, surface and aloft, in figures 4.1 and 4.2, are good examples. They lie definitely on the cold side of the zone of overall temperature contrast (figure 4.3) but not in the heart of the coldest air. They are often called *cold lows*. Meanwhile, the southern portion of the upper trough moves on eastward, perhaps to find another location favorable for development of a new surface low.

Frontogenesis— the birth of a front

The examples in figures 5.1 and 5.3 are ones in which the orientation of the isotherms changes, but not their spacing. That is, the strength of the temperature gradient does not change appreciably. With the usual checkerboard pattern of alternating lows and highs, however, there will inevitably be regions where the geostrophic flow pattern as shown by the isobars will be acting to crowd the isotherms together in a way that is characteristic of a front. An example of such a pattern of *frontogenesis* is shown in figure 5.5. Notice how the isobars show northerly wind in the cold air and southerly wind in the warm air, so the tendency is for the isotherms to become closely packed. The principle works the same as in the other examples: we go along the isotherms with the cold air to our left. But in this case the geostrophic wind changes from easterly to westerly as we go across the saddle point, or *col*, in the pattern of isobars. Now the geostrophic change vector is toward the east, and the low-level nongeostrophic wind component is toward the south (figure 5.6), from colder to warmer air. Hence the warm air ascends and the cold air sinks, and adiabatic temperature changes above the surface layer tend to prevent the isotherms from becoming as close together as the geostrophic flow pattern would indicate.

Since this pattern may extend for a considerable distance along the isotherms (compare the idealized example in figure

Figure 5.5. *How to deduce the low-level nongeostrophic wind when a front is forming. The solid lines are isobars and the dashed lines are isotherms, heavy for an earlier time and light for a later one.*

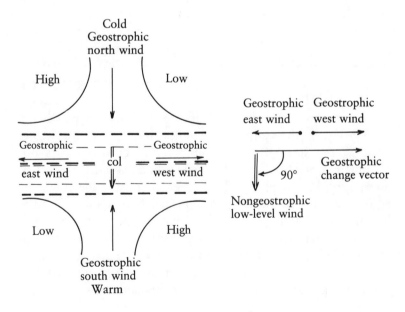

5.6 with the actual frontal trough in the southeastern portion of figure 4.1), an elongated frontal band of cloud will tend to develop in the warm air. In this case, it is worth looking further at what happens near the ground. The vertical motions here are very weak compared with their strength aloft, since strong updrafts occur above and not within a layer of strong near-surface convergence. Therefore there is little possibility of adiabatic warming and cooling near the surface to counteract the tendency of the horizontal flow to bring the isotherms together. In fact, near the warm edge of the nongeostrophic flow this component of the wind adds to the geostrophic effect of compressing the isotherms and a real frontal discontinuity can develop along the isotherms in a few hours, as the cold air wedges in beneath the rising warm air. The wind-shift and temperature change experienced at sea, for example, from warm southwesterly to cool northwesterly, can then be almost instantaneous.

Figure 5.6. *Inferring the weather that will be associated with the birth of a front.*

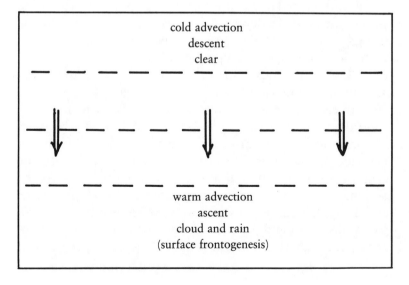

cold advection
descent
clear

warm advection
ascent
cloud and rain
(surface frontogenesis)

Frontal types and frontal cyclones

Fronts are characterized according to how they are moving, as we saw in Chapter 1. The usual situation is for the colder air to be advancing more or less rapidly, in effect sweeping the warm air aloft out of its way. These are cold fronts. Sometimes the cold air is doing little more than holding its ground and may be giving way begrudgingly before a strong onrush of warm air. These are warm fronts, and are usually weakening or dissipating. If the cold air is flowing vigorously more or less parallel to a front in which the warm air is riding aloft over it, we have (not surprisingly) a stationary front. When the cold air has swept far enough around a vigorous surface cyclone that the original warm air can no longer be found at the center, the residual cold front may be denoted an occluded front.

Many surface lows form along preexisting fronts or develop simultaneously with fronts. Figure 5.1 is then altered only in that the isotherms are crowded together at the warm edge of the zone of contrast, where the low also is found (figure 1.1). The isobars around it would be more lens-shaped than circular, elongated in the direction of the front and located just west of the crest of a developing warm ridge. As the developing trough and ridge in the isotherms become more prominent with the approach of the upper cold trough, the frontal low, like all lows, migrates across the isotherms toward the colder air and the frontal cloud band follows it (figure 1.2) along the western edge of the warm ridge. Subsequently the temperature contrast across this part of the cloud band usually becomes insignificant, since the original warm air is no longer found on one side of the front. The front is now occluded, with the original warm air found, if at all, only at upper levels. As the cyclonic circulation becomes more intense both at the surface and aloft, the surface discontinuity of temperature is completely dissolved and the sharp frontal wind shift almost completely so. Only the occluded frontal cloud band persists, out on the periphery of the cyclone, perhaps wrapping more than once around the circulation center (figure 1.3).

Even when the history of a cyclone does not involve a front in the early stages, but rather a broad zone of temperature contrast, the cloud pattern will tend to develop in this way. The vortex outlined by the banded cloudiness is a ubiquitous feature of cyclones at sea. A number of examples can be seen in figures 1.1–1.3. A strongly developed system, having achieved this mature state, may continue to whirl for several days over the ocean, moving slowly if at all, gradually losing strength until it disappears by absorption into the approaching circulation of a newer and more vigorous cyclone.

6

The Message of the Clouds

Clouds show up a great variety of events in the atmosphere, some of them involving changes in the wind. In pictures taken from satellites the clouds map out the large-scale weather systems, particularly the lows with their fronts and other troughs, and they suggest many varied smaller-scale wind movements down to the limits of resolution of the pictures.

Looked at from below, the clouds are just as meaningful, particularly if we know something about the large-scale patterns. The sequence of clouds ahead of a cyclone tells of its advance (see Chapters 1 and 5). But the majority of cloud messages relate to events on a much smaller scale, down to a few hundred yards or so. Cumulus clouds, for example, tell of pockets of rising air, which are replaced by air moving downward between the clouds.

Every cloud has a message of some sort, although not necessarily about the wind. The sailor needs to recognize features that say something about the character of the wind or about changes in the general weather picture, which in turn imply major changes in wind speed and direction. In general the message of the high clouds is about events far aloft or at least several hours ahead, while low clouds provide evidence of wind changes from minutes to two or three hours ahead. We must also try to rule out signs that are unimportant or possibly misleading.

Black, white, and colored clouds

The color of a cloud depends on how it is illuminated. If the sun is shining on it, the cloud will appear white; if the sun is behind it, it will appear dark. If it is illuminated at a glancing angle when the sun is rising or setting, it will be beautifully colored. The color will change as the cloud moves across the sky and the sun moves toward the zenith or toward the horizon. This change of color normally has no significance where the wind is concerned, except that enshrined in the well-known adage "red sky in the morning, sailors take warning; red sky at night, sailors' delight." This saying is well founded. Some of the most colorful skies at sunrise are when the high clouds are increasing from the west and are illuminated at a glancing angle by the sun—and we know that this situation is characteristic of an advancing low, with bad weather to come. Conversely, clouds breaking up from the west are often the last evidence of a weather system moving away to the east, and illuminated by the setting sun these too appear beautifully colored. The bad weather has gone by, and is likely to be followed by an anticyclone or a ridge of high pressure.

Names of clouds

Clouds are named and classified according to their height and shape (see photographs and figure 6.1). Flat clouds are defined by the word *stratus* (meaning "spread out") or by the prefix *strato-*. Lumpy or bulging clouds are described by the word *cumulus* (meaning "heap") or the prefix *cumulo-*. Feathery clouds high in the atmosphere are known as *cirrus* (meaning "curl" or "tuft") and the prefix *cirro-* is used to describe other clouds at high levels. The prefix *alto-* identifies clouds in the middle levels of the atmosphere. The words *stratus* and *cumulus* alone identify clouds whose base is below 7000 feet. *Stra-*

tocumulus is a layer of low cloud which has lumpy or roll features. *Nimbus* is used to describe a raining cloud. *Cumulus* may be combined with *nimbus* to give *cumulonimbus* — heaped rain cloud. A layered cloud from which rain is falling is *nimbostratus*.

Flat clouds

Flat clouds are characteristic of stable air. They often possess some shape or structure, but this is usually due to heating or cooling of the top of the cloud and must not be interpreted in terms of any wind pattern at the surface. However, a large and distinctive feature appearing in a layer of cloud may have some significance.

A line or band of low cloud, or thicker cloud within a cloud layer, may indicate a change in wind speed or direction or both. If the line or band is stationary it may be due to local convergence of airstreams caused by a feature of the nearby land or a significant change in water temperature over the sea.

Figure 6.1.

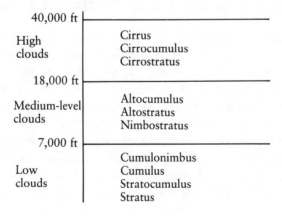

40,000 ft	
High clouds	Cirrus Cirrocumulus Cirrostratus
18,000 ft	
Medium-level clouds	Altocumulus Altostratus Nimbostratus
7,000 ft	
Low clouds	Cumulonimbus Cumulus Stratocumulus Stratus

Cirrus.

Cirrus increasing ahead of a warm front, cumulus below.

Stratus.

Stratocumulus.

If the line is moving it is clearly a feature of the air mass and a small windshift (normally a veer) is likely as it passes by. Think of these features in the atmosphere as small-scale replicas of the weather-map systems: An advancing line of cloud can be thought of as a very minor trough giving a small and probably temporary veer in the wind direction as it goes by.

If there is little or no wind, an approaching line of stratocumulus cloud is likely to signify wind from a direction to the left of the line of advance; in other words, the line of advance indicates the direction of an increasing geostrophic wind, with the surface wind some 20 degrees to the left of it.

There are many variations on the theme of lines and bands of cloud, and it is important to realize that the atmosphere is never completely uniform over an area, even over a smooth sea. There are always irregularities, and it is not easy, even for a meteorologist, to distinguish between those lines of cloud that suggest a wind change at the surface and those that do not.

Lumpy clouds

Lumpy or cumulus clouds are characteristic of unstable air. They are found most frequently over land in the afternoon with the temperature at a maximum, when pockets of air heated at the ground rise until the cooling due to expansion brings their temperature back to that of the surrounding air.

Look for a moment at a single cumulus cloud. If it is stationary—that is, if there is no overall wind—air will rise from the heated surface, its temperature falling at the adiabatic lapse rate (see Chapter 2) until it reaches its dew point. Condensation will then occur and cloud will form. The nearer the dew point is to the starting temperature, the less the air has to cool before cloud forms and the lower is the base of the cumulus cloud; and vice versa. Above the cloud base the now cloudy air will continue to rise as long as it is warmer than the air around, and its temperature will fall off at the lesser moist adiabatic lapse rate. A hard-edged cauliflower-like top to the cloud indicates air rising strongly in a vigorously growing cloud. A fluffy top indicates that the air has virtually stopped rising. The size of the cloud provides a good guide to the wind strength in the area of inflow below the base of the cloud. For a small cloud, say about 100 yards across and a thousand feet in vertical extent, it will be less than a knot; but for a towering cumulus cloud on the order of 1000 yards across and up to 15,000 feet deep, it may be six or so knots.

Actually there is usually some wind moving the cloud along, and the gentle convergence of air into a small cloud is only a minor detail superimposed on the main wind.

Sometimes cumulus cloud is seen to develop strongly to start with, then flatten out into a layer. This indicates that the rising air has stopped suddenly on reaching a temperature inversion, a feature often associated with an advancing ridge of high pressure.

Figure 6.2. *Formation of a cumulus cloud.*

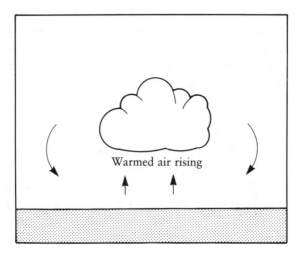

Warmed air rising

Cloud streets

Over the sea cumulus clouds are normally found in regularly spaced lines. (See also Chapter 10.) The best examples of these are found in the trade winds, where they extend for many hundreds of miles. Cloud streets show up in a regular pattern in the vertical movement of air, which is like a horizontal roll. Between the lines of cloud you will find the stronger, more gusty and slightly veered winds, and beneath the lines of cloud somewhat lighter and more backed winds (figure 6.3).

Sometimes you may notice near land a single cloud street, a single line of cumulus clouds downwind from something that

Figure 6.3. *Cloud streets.*

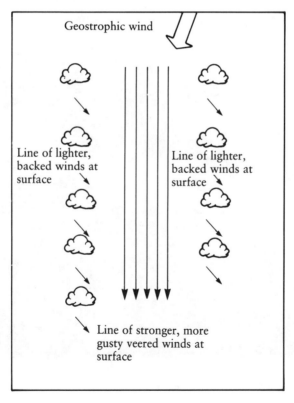

Fair-weather cumulus.

A line of cumulus or "cloud street."

The top of this cumulus is flattening into stratocumulus—typical of a ridge approaching.

Altocumulus.

is producing them—a hot spot, a hill or an island. Once produced, the cumulus clouds are carried away in a line stretching
downwind for many miles, and the pattern of lighter and
backed wind under the street, with stronger and usually veered
wind either side of it, will also persist.

Raining clouds

So far we have considered fair-weather clouds, or at least
clouds giving no evidence of rain falling. Rain makes a fundamental difference in the wind characteristics of a cumulus
cloud. The main reason for this is that the first rain to fall from
the base of the cloud evaporates into the air beneath and cools
it, often by several degrees. This cooled air descends, and the
more it is cooled the more rapidly it descends. Thus, instead of
air rising into a cloud we have not only rain falling out of the
cloud but air as well. The drier the air beneath the cloud the
more readily it may be cooled by evaporation, and as long as
there is enough rain coming out of the cloud the colder the air
becomes. The cooled air will literally drop from beneath the

Figure 6.4. *Cooled air falling from a raining cloud.*

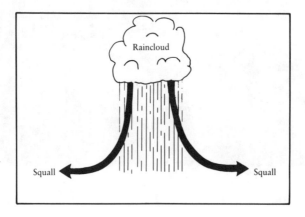

Large cumulus approaching cumulonimbus stage.

Cumulonimbus.

cloud with the rain and spread out in all directions at the surface. The light wind which was moving in toward the cloud suddenly becomes a squall rushing out and away from it (figure 6.4).

There is always plenty of visual evidence of this change. You can see the rain falling, often in gray streaks below the cloud, sometimes in a dramatic arch of black cloud spreading out from the parent cloud. Most such squalls are short-lived because there is only a limited amount of air below the cloud to be cooled by the evaporation of rain into it. Once the squall has passed the rain usually continues for a while, some ten to twenty minutes for a typical shower cloud before it is exhausted. The wind coming out of the cloud will gradually die away with the rain. Usually there is an overall wind, but often the squall from a reasonably sized shower cloud will temporarily override the large-scale wind, and will augment it where they are in the same direction. Conditions associated with thunderstorms and the more severe squalls are discussed in Chapter 15.

You may be wondering about the saying, "There is always more wind under the cloud." This applies only in the case of calm or very light winds when there are large cumulus and cumulonimbus clouds around. You then have to find the wind caused by the updraft into the cloud or the squall coming out of it when it is raining.

Cloud bands

There are many different kinds of cloud band, and they form for a wide variety of reasons. Most of them tell us something about the wind. Examples are bands of thicker (or thinner) cloud within a layer of cloud in stable air, lines of cumulus clouds, cloud streets, and so on. A band of cloud may be observed lying along the coast, layered if the air is stable or cumulus if the air is unstable. If the coast is fairly flat this band is likely to indicate a convergence of airstreams. A hilly (and particularly a mountainous) coast will frequently appear cloudy simply because the air has to rise over the high ground irrespective of wind direction.

An approaching band of towering cumulus clouds is usually associated with a trough of low pressure, quite a frequent occurrence in an unstable air mass. The intensity of the trough can vary greatly: the cloud and rain may pass in ten minutes or take several hours. In a satellite picture a minor trough is seen as a fairly narrow band of cumulus cloud. On a weather chart the plotted observations will show falling pressure ahead of the trough and rising pressure behind it, with stations situated near the line of the trough reporting a shower. In terms of wind, a backing of five or ten degrees might be observed 5 to 30 miles ahead of the trough, and a veering of about the same amount as it passes.

Some of the most interesting features are the long bands of high cloud stretching from horizon to horizon with high-

Castellated altocumulus often heralds thundery weather.

level winds blowing along them, sometimes at speeds of 150 knots or more. Such clouds are characteristic of a high-level jet stream and can be very useful predictors (see photograph). If the band of high clouds is stationary the weather is unlikely to change in the next twelve hours or so. If the band is the forward edge of increasing or thickening high cloud, it tells of an advancing cyclone or trough, perhaps a front. Conversely, if the high cloud is moving away and the sky is clearing, it suggests that the system has passed and better weather is on the way.

These conclusions can be made more certain by noting, if possible, the direction of movement of high cloud. If the wind aloft in an advancing area of high cloud is markedly veered from the surface wind (say 90 degrees or more), it is a certain sign of an approaching trough, perhaps a warm front. On the other hand, observations of the high cloud associated with a receding cold front indicate upper winds that are markedly backed from the direction of the surface wind. We saw in Chapter 3 how the vector difference between the surface wind and the upper wind—the thermal wind—tells you where the cold and warm air are. Cold air is to the left and warm to the right as you stand with your back to this thermal wind.

A band of high cloud—cirrocumulus and altocumulus—a useful predictor of weather to come.

Clouds of a cyclone

We have seen that the ascent of air in a low is concentrated on its forward side, and it is no surprise to find the thickest cloud and heaviest rain there. Being on the forward side of the low, the clouds herald its approach, and typically a sequence of thickening cloud at high, middle, and finally low levels is observed as illustrated in figure 6.5. The various forms of cirrus cloud at high levels display an often beautiful variety of shapes and forms depending on the wind and humidity aloft. The two things to notice are the speed at which the clouds increase and the speed of the winds aloft. A rapid increase in the clouds at all levels at once, with little indication of wind shear, suggests the approach of a relatively innocuous trough which may soon pass. A substantial increase in cloud starting at high levels and gradually working downward, with indications aloft of strong winds well veered from the surface wind, hints at the approach of a deepening low with attendant strong winds. The movement of your barometer will help you to confirm predictions from the clouds.

A front is a special case of a trough of low pressure. The cloud sequence ahead of a warm front is similar to that ahead of a low, the main difference being the stronger shear aloft due

Figure 6.5. *Vertical section of clouds ahead of a low. If a warm front is present, it will lie along the dashed lines.*

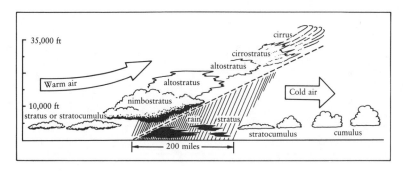

to crowding together of the isotherms (to give a stronger thermal wind). In the warm air behind the front, the clouds will be of the stratus type if the water is cold, typical of warm, moist air being cooled from below. The closer the dew point is to the sea temperature the lower will be the cloud, and indeed, widespread dense fog is often found when the sea is cold enough. Aloft, especially when the sea temperatures are high, towering cumulus clouds can often be seen, reminding us of the tropical origin of the air.

A cold front normally features cumulonimbus clouds with their tops spreading out in the strong wind shear aloft. In contrast to the warm front, however, the shear-revealing clouds can only be viewed from behind the front, after it has passed. The greater the contrast between the warm and cold air at the surface, the greater the likelihood of thunderstorms and severe squalls.

7

Moving Weather

The practical application to sailing of what we have discussed so far means becoming familiar not only with signs of changes in the weather and the general workings of weather systems, but also with the simple geometry of changes in wind direction and speed as lows and highs, troughs and ridges pass by. For the racing sailor the outcome of a race will normally depend critically on whether these changes have been correctly anticipated. But this anticipation is equally relevant to cruising. Using a variable wind to advantage can save many hours on an ocean passage, and the resulting sense of achievement is not to be underestimated.

The terms "veer" and "back" have already been introduced, but it is appropriate here to include definitions and also to underline their relevance to sailing. The wind is said to "veer" if it swings in a clockwise direction, and to "back" if it changes in a counterclockwise direction. These terms are used to describe changes experienced in time as well as changes in the vertical or along the horizontal. This is illustrated in figure 7.1. A veering wind is therefore a "header" if you are sailing on port tack, and it provides a "lift" if you are on starboard tack. Conversely, a backing wind is a header on starboard and provides a lift on port tack.

What happens to the wind as a sequence of highs and lows or troughs and ridges moves by? Let's look at some typical examples. We will assume you are in the open ocean well away from complications due to coastal effects and land or sea breezes, moving slowly compared with the 20 knots at which these weather systems typically progress in summer. The sequence is usually similar to the sequence along a line passing through the system, parallel to its track across the map. Ac-

Figure 7.1.

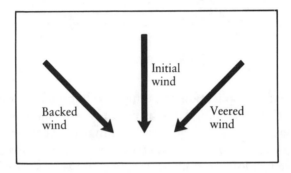

Figure 7.2.

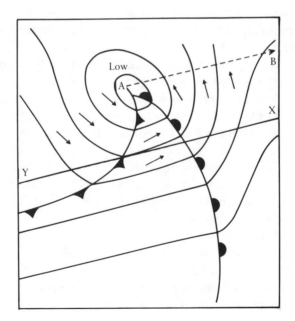

count has to be taken of the changing orientation of the troughs, ridges, and fronts that may rotate around large low and high centers.

A low moves east to the north of you

AB is the track of the low center (figure 7.2). As it moves toward B the sequence of wind direction and speed at your position X will be similar to that along the line XY. The wind will increase and back toward south-southwest, veer slightly as the warm-front trough passes, then remain steady until the sharp veer to northwesterly at the cold front—about 70 degrees in this example. Presently the wind will weaken and back toward westerly as the high pressure ridge approaches. Often there will be no perceptible warm-front shift.

Clouds will thicken and there may be a little light rain when the wind is south-southwest. Improvement occurs, except for the possibility of fog, as the wind veers to southwest. A brief period of showers, which may be heavy, precedes or accompanies the cold-front passage. Cumulus clouds, and perhaps scattered showers, are the rule in the northwesterly flow following the cold front, unless land is to windward, as along the east coast of North America. These weaken and die out as the high-pressure center approaches.

A low moves east to the south of you

AB is the track of another low center (figure 7.3). It is moving east, but this time to the south of your position at X. The wind backs from west-southwest at the start through east to northerly at Y. There are rarely any surface fronts along the line XY. Only an occluded front may be found in these sectors of the periphery of an old mature cyclone. In this case a strong easterly may be replaced temporarily by a weaker southeasterly as the occlusion rotates around the low center.

Despite lack of surface fronts, there is plenty of weather. A substantial period of overcast clouds and rain, at least 12 hours long, is experienced in this situation, since the ascent of warm, moist air is greater and more prolonged to the left of the cyclone track than to the right. The high clouds typically on the increase when the wind first backs to east subsequently thicken and lower, with rain soon to start. The wind must back to north and then northwest before improvement can be expected. Of course, if the track of the low is far enough away, say, no closer than about 300 miles, then you may escape its consequences altogether and see only the fringe of high cirrus clouds to the south.

Figure 7.3.

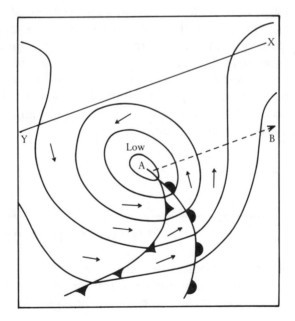

_____ A ridge moves across you

The ridge with axis H-H moves eastward (figure 7.4). Your wind at X will be a steady northwesterly at first, but then will back to west and decrease gradually as pressure rises to a maximum. Winds near the axis H-H will be very light, but then will increase again as the direction continues to back toward south and the pressure begins to fall with the advent of the next low. The sun you may enjoy by day, and the stars and moon by night, when the high is in the vicinity, will become obscured by the high clouds of the oncoming cyclone. If the high center itself is at the latitude of 30° to 35° N it may be large and nearly stationary; the Azores and Eastern Pacific highs are examples. The next cyclone or low-pressure trough may be several days away! It's time to use some fuel.

Figure 7.4.

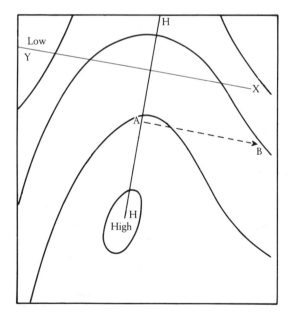

Other possibilities

In these simple examples, the lows and highs are taken to be moving directly from west to east. If the track of the center is in some other direction, then figures 7.2 through 7.4 have to be rotated accordingly, with the wind directions consistently altered.

For example, if a low is moving from southwest to northeast, as it typically is, then the wind will be first southeast, then south, and finally westerly. Occasionally a low may move from south-southeast toward the north-northwest. In this case, an initial east wind will veer to southeast in the warm sector and then into the southwest as colder air arrives from an unusual direction. You can work out other possibilities in each of these figures.

If the low should happen to pass directly overhead, then the wind direction will be nearly constant as it approaches and as it recedes, but you will experience a shift of 180 degrees as the center passes. The weather is intermediate between the two scenarios accompanying figures 7.2 and 7.3, neither as benign as when you are on the right-hand side of the track nor as unpleasant as when you are on the left.

Figure 7.5.

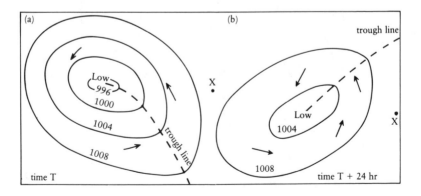

As another example, if the center of the high in figure 7.4 passes eastward to the north of your position, then a northerly wind will veer to northeasterly and then to easterly and perhaps southeasterly. If you are in subtropical latitudes, the weather will remain fair, and trade winds will freshen for a day or more. In middle latitudes, however, the pleasant weather may not last long, since a cyclone is likely to develop and approach from the southwest by the next day.

_____ A stationary filling low with a trough moving around it

The low center remains in the same position relative to yours at X (figure 7.5), but the trough rotates around it in a counterclockwise direction. Although the wind shift at the trough is from south-southeast to west-southwest (a) at the start, it may be from east-northeast to south-southeast (b) by the time the trough reaches you, as in the example of the old occluded front mentioned before. The winds, of course, will decrease slowly as the low weakens.

8

Winds Near Coasts

Winds blowing off the shore

If the wind is blowing from the land, whatever the angle of the wind to the shoreline, the direction of the wind will usually veer and the speed increase as the air moves out over the water (figure 8.1). We have already seen that for a given pressure gradient the difference in the direction of the surface wind over land and over water is about 25 to 30 degrees. This means that you can expect to be headed if you approach land close-hauled on starboard tack. An exception occurs over cold waters when there is a strong offshore geostrophic wind and sunny weather. During the daytime the wind is strong and gusty on land and along the immediate shore, because of downward mixing of strong wind from aloft. A few miles to seaward, however, the cooling of the air by the cold water reduces this mixing and the wind is slightly backed and much steadier, although weaker. Clearing northwesterlies in summer off the New England coast are a good example of this effect.

——— Wind blowing along the shore

You may not have noticed it, but there is a quite remarkable difference in a wind blowing along the coast, depending on whether as you stand with your back to the wind the land is on your left- or right-hand side.

If the coast is on your right-hand side, the different angles of the surface winds over land and water are convergent for the same pressure gradient (figure 8.2). If the air is vertically stable and resists being lifted, this convergence results in a band of stronger wind within two or three miles from the shore. The increase of speed is normally of the order of 25 percent, or 5 knots added to a 20-knot wind well offshore. Some of the air will be forced upward as well, often giving an increase in cloud or a band of thicker cloud along the coast.

The stronger wind just offshore in this case may be mistaken for a sea breeze, but it persists night and day. So if, for

Figure 8.1. *Wind veering over the water.*

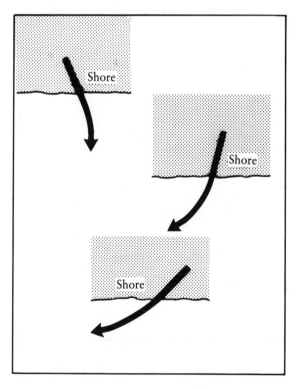

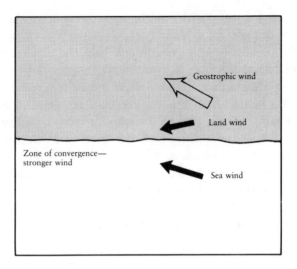

Figure 8.2. *Convergence.*

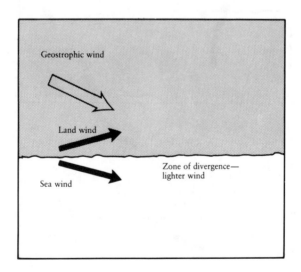

Figure 8.3. *Divergence.*

example, you leave harbor on an east-facing coast in a southerly wind and find it 20 knots instead of the expected 15, you can be reasonably confident that by the time you are three or four miles off the coast it will have dropped to the lower speed you were hoping for. Equally, as you approach a coast when the winds over land and sea are converging, be prepared for an increase. Along an undulating coast with bays and headlands not more than about 15 miles apart, the band of stronger winds is found along a line joining the headlands, and you have to sail three or four miles seaward of the headlands before reaching the lighter wind.

If the air is vertically neutral or unstable, then the converged air goes mainly up rather than faster along the shore. In this case a line of clouds will tend to occur near the shore, over land when it is being strongly heated or over water when the land is being cooled and the water is relatively warm.

When, as you stand with the wind at your back, the coast is on your left-hand side, the effect is opposite (figure 8.3). The airstreams diverge, the wind is that much lighter near the shore, and there is less cloud than elsewhere. Note however that in the afternoon this lack of wind near the coast is often overridden by a sea breeze.

Wind blowing onto the shore

There is no significant variation in the wind over the water. All of the changes occur over land.

Coastal cliffs

If the wind is blowing *along* the shore it makes little difference whether there are cliffs or not; the effects described above are well marked and sometimes enhanced. That is, we observe within a few miles of the coast a zone of stronger or lighter winds, depending on the wind direction.

If the wind is blowing *off* the shore, it will be affected by the cliffs (figure 8.4). Standing waves often form in the wind downwind of the cliff face and give relatively static zones of stronger and lighter wind, sometimes marked by a cloud sitting on top of the lighter wind zones. The zones of stronger wind are the more reliable and are likely to remain in the same place for as long as the wind direction and stability of the airstream do not change. The zones of lighter wind may be characterized by considerable variations, even reversals in wind direction, particularly downwind from the higher cliffs, but the positions of the zones themselves are likely to stay put for some time. Beneath the cliff itself there is usually a large eddy in the flow of air with a complete reversal in wind direction. Calms, reversals, and blasts from the prevailing direction make anchoring a real problem, contrary to one's expectation of shelter in the lee of an orographic obstacle.

The west coast of North America abounds in examples of these effects, from Mexico to the United States to British Columbia and southern Alaska. In contrast, land elevations are low and slopes are gentle along the East Coast, except in parts of Maine and the Canadian Maritime Provinces.

Figure 8.4. *Coastal cliffs.*

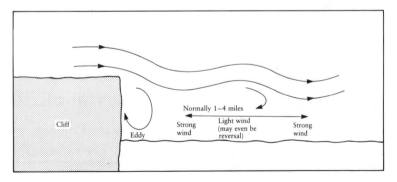

An island

A mountainous island clearly obstructs the wind, and not surprisingly its influence extends many miles downwind. The islands off southern California and Baja California are good examples. Downwind from these islands large eddies form in the air flow, often between fifty and one hundred miles across. Having formed, they move away downwind, weakening only slowly. Satellite pictures may show a string of three or four eddies, the farthest being some two or three hundred miles from the islands.

Even a small, flat island has a significant influence on the wind up to five miles or more to leeward. The air flowing over

it is subject to greater friction, so it slows down and its direction backs some 15 degrees. Along one side of the island a zone of stronger wind will be caused by converging airstreams, and along the other side there will be a zone of lighter wind caused by the diverging airstreams (figure 8.5).

These zones of stronger and lighter wind will not be limited to the island shore but will continue downwind for a considerable distance, perhaps tens of miles. Sometimes a line of cloud or individual clouds will provide evidence of the position of the stronger wind band, and a band of blue sky or thinner cloud, evidence of the lighter wind band.

Figure 8.5.

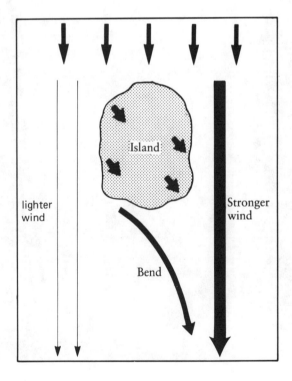

Land breezes

These are found near coasts at night or in the early morning, usually when the sky is clear. The best way to visualize them is as "drainage" winds. Air that has cooled over the land on a clear night, being relatively dense, drains downward, usually following valleys, until it reaches the sea. Its momentum carries it a mile or two out to sea before it warms up and dies away.

The direction of a land breeze is controlled almost entirely by the contours of the land. The cold air flows down the valleys and on reaching the sea spreads out fanwise. The steeper the slope, the stronger the breeze. Light land breezes can be found close to a coast that is low and flat, even when the contours can barely be distinguished.

An offshore geostrophic wind may be enhanced by a drainage land breeze as described above. In the absence of a drainage wind, however, an offshore wind at night may be modified by another effect. The land cools at night much more than the sea. Because of the strong stability that develops, any wind at night near the land surface is likely to back and decrease, and may even die away altogether. In contrast, it continues almost unchanged over the open water well away from shore, as we saw in figure 3.4. Thus, near the coast at night there will be a zone where the wind is picking up gradually toward its open-sea speed and direction.

Mountain and valley winds

In mountainous areas the geostrophic wind can either be bent to blow along the valleys, if it is blowing somewhere near that direction, or else it will blow across the top of the mountains, leaving only eddies or stagnant air in the valleys.

In addition there is frequently some form of drainage wind to be found—either as a result of cold air accumulating in the valleys or because the cold air mass accompanying a high-pressure area piles up behind a mountain barrier until it finds a way out. Then a "gap wind" can burst through, sometimes violently. Both types of drainage wind can arise very suddenly and both show some diurnal variation. They are strongest in the early morning when the air is coldest and lightest in the afternoon when the air is warmest. If the cold air is simply the result of nocturnal cooling, the drainage wind will cease completely during the day when the land warms up, to be replaced by a wind blowing up the valleys toward the surrounding hills.

On the west coasts of Central and North America prominent mountains come close to the shore. The piling up of cold air east of the mountains can produce spectacular examples of gap winds. Cold air over the Gulf of Mexico pours through the Isthmus of Tehuantepec in Mexico, as a violent wind called the "Tehuantepecer" on the Pacific side. Its influence can be felt more than one hundred miles from shore. On a smaller scale, the "Santa Ana" of southern California is a strong and gusty northeasterly which can endanger small craft. Coastal waters near the mouth of the Columbia River in Oregon are reached by the extension of damaging winds that sometimes funnel through the Columbia Gorge in the winter, while many locations along the coasts of British Columbia and southern Alaska feel the effects of gap winds that can reach hurricane force.

Another powerful effect on winds in North America is indirectly due to the major mountain ranges. When the northeasterly flow of cold air in the southern quadrant of an anticyclone is blocked by the east-facing slopes of the Appalachians or the major ranges of the western part of the continent, the air can go up or it can go south. Since this air is usually very stable it resists lifting and flows southward along the mountain slopes and along the lower land to the east, often with great strength. In the West, this flood of cold air down the Plains is called a "norther." This wind can produce dangerous conditions in the Gulf of Mexico and into Central America. In fact, the Tehuantepecer is part of the norther.

Along the East Coast, the cold front at the leading edge of this more modest flow of cold air moves southwestward, often surprisingly rapidly. Because it moves from a direction other than west, from which most weather systems come, it is called a "back-door cold front." It is rarely welcome, because it signals the end of a spell of pleasant, warm summer weather, sometimes with thunderstorms and sharp squalls. Should the warmth have developed into searing heat, however, the back-door cold front brings in a refreshing breeze with a marine origin. When, especially during the cold season, this flow of air enters the leading edge of a rapidly developing low moving northeastward along the coast, then it can strengthen to dangerous proportions, and may become the infamous "nor'easter" of New England.

9

The Sea Breeze

The sea breeze is the result of air being heated over the land more than over the sea, and this can happen not only on a sunny day but also when there is thin or patchy cloud. The first question to ask is, "Will the air be heated strongly over the land during the daytime?" If so, the sea breeze mechanism will begin to get underway as follows (figure 9.1):

1. The air over the land is heated and expands.
2. This causes an excess of air (and hence a small high-pressure area) at some higher level, usually 1,000 to 3,000 feet, over the land.
3. Air flows seaward aloft in response to the pressure differential from land to sea.
4. This air subsides slowly over the sea.
5. At the surface, the pressure drops over land and rises over the sea, because of the transfer of air aloft, and in response the air at the surface begins to move across the shore—which is the sea breeze.

Actually, these steps follow each other in the matter of minutes, so they can be thought of as happening simultaneously. This simple case assumes no overall geostrophic wind.

As time goes on the sea breeze circulation extends steadily both inland and seaward, and the direction gradually veers (turns clockwise). Because the breeze experiences greater drag as it moves over the land, it tries to realign itself to a direction more parallel with the coast. The earth's rotation does not cause the realignment, as many books state, but its influence is just enough to flick the breeze to the right (northern hemisphere) or to the left (southern hemisphere).

The vertical extent of the breeze depends on the stability of the air. In stable air the return flow may be confined below 2000 feet or so, in which case there is only limited scope for the sea breeze to develop, and it is likely to be weak.

For the sailor, in cloudy conditions it is the downward movement (subsidence) of the air over the sea feeding the sea breeze that is the most important feature. One of the first clues

Figure 9.1. *The sea breeze.*

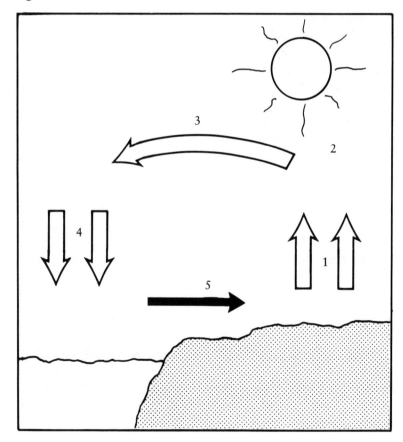

to the beginning of a sea breeze is often the fairly sudden dissolving of low cloud just offshore, a sign of air gently subsiding in preparation for feeding the first drift of air onto the shore. When the sky is clear, the first sign may be the sudden development of a line of small cumulus clouds just inland from the shoreline, visible evidence of the ascent of heated air over land.

On a straight coastline, therefore, and in the absence of any overall pressure gradient, the observable steps in the development of a sea breeze are as follows:

1. Calm morning, clear sky or thin cloud. Temperature over land begins to rise.
2. Cumulus cloud begins to form over land, or cloud begins to disperse just offshore.
3. Gentle drift of air onto the coast starts within about a quarter to a half mile offshore.
4. Breeze gradually increases and extends seaward.
5. Direction of breeze gradually veers, and its strength increases to 10 to 15 knots about two or three hours after the start. The strength depends on the stability of the air and the heating contrast between land and sea.
6. By late afternoon the direction has veered to about 20 degrees from the orientation of the shoreline, a shift of 50 to 60 degrees from the start, and the breeze extends up to 10 or 15 miles seaward.
7. The sea breeze dies away toward sunset, depending on the rate of fall of temperature over the land. The dying remains of the sea-breeze circulation may continue to drift seaward and may be found on occasion up to 30 or 40 miles from shore in the middle of the night.

Complicated coastlines

A large bay with a narrow opening. At first the sea breeze is a drift of air onto all shores of the bay (figure 9.2), but these onshore breezes have difficulty finding enough air to feed them and their strength pulsates between calm and two or three knots. Some air, however, is pulled through the entrance to the bay, and as the breezes keep trying to blow onto the shore so the amount of air pulled through the entrance increases. At the same time the sea breeze develops on shores adjacent to the bay amd eventually, if the bay is reasonably small, takes over the whole area. It will usually be found that the right-hand shore of

Figure 9.2.

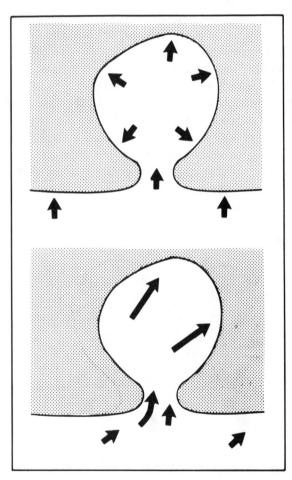

the bay has more wind, and the breeze through the entrance is likely to be somewhat stronger than elsewhere, with a tendency for convergence of air into the entrance and a bend in the wind, particularly on the right-hand side.

These features were researched in some detail in Narragansett Bay prior to and during the 1980 and 1983 America's Cup races. Buzzards Bay does not have an especially narrow mouth, but the afternoon southwesterly, driven mainly by sea breeze effects, can build to more than 25 knots when the initial wind is not strongly opposed. On the West Coast, San Francisco Bay provides ideal opportunities for local sea breezes to be overridden by a breeze generated along the coast on either side. A near-gale from the west funnels in through the Golden Gate during the afternoon.

An island.   In contrast to the bay, here there is plenty of air to feed the sea breeze but, if the island is small, hardly anywhere for the air to go (figure 9.3). In the case of an island on the order of ten miles in diameter, the first gentle onshore sea breeze, five knots at best, will quickly cool the land on which it relies for heat, and will therefore die away to be followed by successive attempts at generation of a breeze at inter-

Figure 9.3.

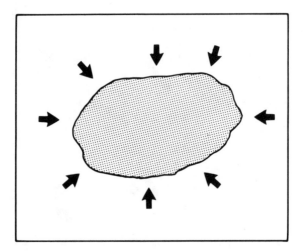

vals of an hour or so. The larger the island the more successful will be the attempts at a breeze. If the island were perfectly flat and uniform there would be a gentle onshore breeze all around, and cloud would likely be seen sitting over the middle from time to time. In practice the sea breezes are critically dependent on the topography, and even small hills, bays, and valleys will cause good local breezes at the expense of calms elsewhere.

An island near the mainland. In this case the mainland sea breeze will gradually take over and eventually swamp the more local breezes on the island, although not before there have been successive stages of light wind and calm over the more enclosed area of water as breezes try to blow onto all shores. A pattern is shown in figure 9.4.

So we see that as a general rule when there is no geostrophic wind, however complicated the coastline, the sea breeze will start blowing gently onshore and then develop into a breeze whose direction largely irons out the minor features of the shoreline, veering slowly until the direction approaches the overall line of the coast.

Sea breeze with initial wind blowing

Let us see how a sea breeze is affected by a wind already blowing. This will normally be the geostrophic wind. We know that two essential components of the sea breeze circulation are:

Figure 9.4. *Sea breeze development over Long Island.*

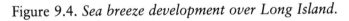

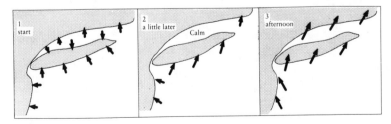

• An opposite (offshore) breeze blowing at some height above
the surface.
• The subsidence of air over the coastal water.

In the absence of a pressure gradient there is no hindrance
or help to either of these components, and we experience a sea
breeze whenever the heating over land raises the temperature
to a value above the sea-surface temperature. But when there is
an appreciable geostrophic wind—and there usually is—the
onset of a sea breeze is critically dependent on whether this
initial wind is blowing offshore and encouraging the upper part
of the sea breeze circulation. This only applies if the geostrophic wind is less than 20 to 25 knots; if it is stronger than
this a sea breeze will never develop, because the geostrophic
wind will reach down to the ground and overwhelm any at-
tempt at development of a local circulation across the coast.

*Figure 9.5. Sea breeze development in the Strait of Juan de
Fuca. The average sustained sea breeze on the Canadian side
(Sheringham Pt., Race Rocks, Trial I.) is shown in the solid
line, and the U.S. side (New Dungeness, Smith I., Pt. Wilson)
is represented by the dashed line. Vertical shaded area shows
nighttime hours; vertical arrows are measured maximum wind
gusts from U.S. stations. (Canadian stations did not measure
wind gusts.) Note that average winds are stronger on Canadian
side and that the sea breeze increases in the afternoon, reaching
a maximum shortly before or after sunset. A 39-knot gust
occurred at 1:00 a.m., 5 August. (Reprinted from* Marine
Weather of Western Washington *by Kenneth E. Lilly, Jr.
Starpath School of Navigation, Seattle, WA)*

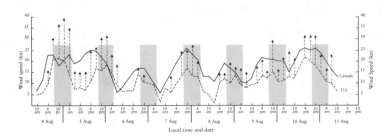

Onshore initial wind

With the wind blowing onshore you never get a true sea breeze, because the return flow aloft is prevented. However, the initial wind is modified as a result of the heating of the land during the day, the heating causing the pressure to fall. The greater the heating the greater the fall of pressure, and therefore the greater the modification of the initial wind. Applying Buys-Ballot's Law we see that onshore winds angled to the coast so that land is to the left (back to the wind) have relatively low pressure over the land. In this case the further fall of pressure due to heating enhances the geostrophic wind with a component parallel to the coast. The result in warmer latitudes such as California is an extra 10 to 15 knots of surface wind by midafternoon. The effect is smaller farther north.

If on the other hand the onshore wind is oriented so that land is to the right (back to wind), the initial pressure is rela-

tively high over the land. Now the reduction in pressure over the land due to heating means a reduction in the pressure gradient, and a decrease in the wind is experienced. The hotter the day the greater the decrease. It may be, however, that the initial pressure gradient is weak to start with, and the heating of the land may be sufficient to neutralize it completely. In this case we have a situation without any pressure gradient at all, so that a pure sea breeze is free to develop. A morning wind of five or six knots from the land dies away slowly as the land heats up, and after a calm period, which may be very brief or may last two to three hours, a good sea breeze springs up just when you had given up on it. Only local experience can teach you how to anticipate such a sequence, but an understanding of the mechanism will provide a framework for your observations.

Offshore initial wind

The development of a sea breeze with an offshore geostrophic wind involves a calm period as the offshore wind dies and makes way for the sea breeze blowing onshore. The behavior of this calm patch depends on the angle of the offshore wind to the shoreline. A divergent zone (Chapter 8) encourages the subsidence that feeds the onshore sea breeze. Typically in this case the wind falls to calm close inshore by midmorning, followed quickly by a sea breeze onto the beach which builds steadily and extends out seaward, while at the same time pushing inland. The calm zone separating the sea breeze from the offshore wind will be found about five miles offshore within an hour or so, and 20 or more miles offshore by midafternoon. If, however, the initial offshore wind has a direction within about 30 degrees from the orientation of the shoreline, a calm zone is unlikely. The onset of the sea breeze merely causes the initial wind to decrease and back by about 60 to 90 degrees, followed by an increase and veer, reaching the final sea breeze speed and direction.

A zone of convergence near the shore opposes the subsidence that is essential to the sea breeze and leads to a more complicated development. The magnitude of this convergence and its influence on the sea breeze depends on the strength of the component of wind parallel to the coast. The sea breeze, instead of starting close to the shore, starts seaward of the zone of convergence, usually three to six miles out, and slowly moves in as it develops, while at the same time extending out seaward. Thus, two calm zones are experienced over the water, with the sea breeze sandwiched between them. A good example of this occurred on Lake Ontario during the 1976 Olympics at Kingston. A northeasterly initial wind (with the coast facing south) persisted over the inshore course throughout the day, while a sea breeze developed seaward of the outermost racing course and moved shoreward across two of the courses, coming to a halt on the Flying Dutchman course, about two miles offshore.

The likelihood of a sea breeze and the behavior of the associated calm zones are summarized in figures 9.6 and 9.7, in which the possible directions of the initial wind are split into four quadrants by a line drawn perpendicular to the coast. It is immaterial which direction the coast itself is facing; the diagrams apply equally to coasts facing north, south, east, or west and all the directions between.

Figure 9.6.

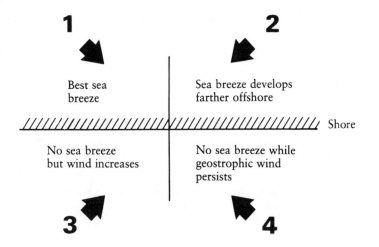

There is just one sea breeze, and it generates the same way the world over. With 360 possible directions of the initial wind and an even greater variety of shapes of coastline, there is scope for a great variation in sea breeze behavior, but it is always capable of being understood as outlined here.

Avoiding the calms

For an initial wind in *quadrant one* the calm period will be short and the best (strongest) afternoon wind will be found within ten miles of the coast, but at other times you should stand farther off.

For a geostrophic wind in *quadrant two* the best night wind will be within ten miles of the coast. If this wind direction suits you then stand farther off during the afternoon and avoid the calm patches. If you want a temporary blow from the opposite direction look for the sea breeze in the zone from two to six miles from the land.

If the initial wind is in *quadrant three* you will get the best afternoon wind within six miles of the shore; otherwise, stand farther off.

The opposite applies in *quadrant four*. Stand in except during the afternoon. If the initial wind is light, look for a late sea breeze.

Figure 9.7.

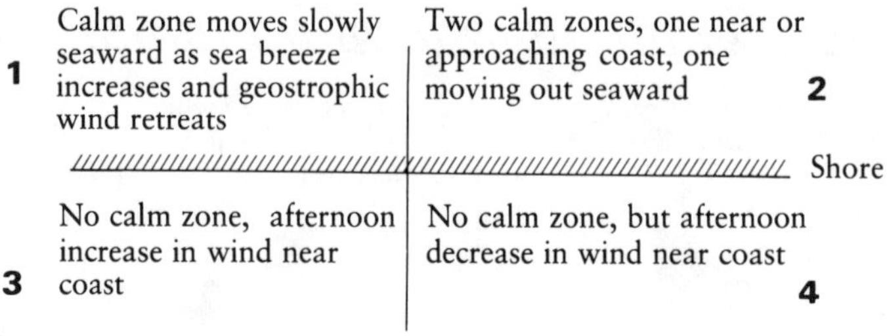

1	Calm zone moves slowly seaward as sea breeze increases and geostrophic wind retreats	Two calm zones, one near or approaching coast, one moving out seaward **2**
	//	Shore
3	No calm zone, afternoon increase in wind near coast	No calm zone, but afternoon decrease in wind near coast **4**

If the coastline is extremely complex

The shoreline is rarely if ever straight, but do not worry too much about the complications unless you are sailing close inshore when the sea breeze is just starting. If you are, you will find that your local initial sea breeze within half a mile from the shore will depend on the orientation of the nearest shore. Thereafter the sea breeze will develop as if to smooth out the variations in the coast, and you can work on the basis of the best straight line you can draw along the coast.

10

Winds Over the Open Ocean

Wind bands

It is a misconception to think of the wind as being uniform for a given pressure gradient. It is not—even over a perfectly smooth sea, if such were possible. The wind likes blowing in bands. The most striking example of this is in the trade winds, which are characterized by lines of "trade wind cumulus" extending hundreds of miles. Examples of this can be seen in figure 1.1 by careful comparison of the fine bands along about 15° N with the isobar in that area. These lines of clouds are evidence of what are called "vortex rolls," wherein superimposed on the horizontal motion of the air is a vertical circulation of air moving slowly up into the lines of cloud and down in the clear lanes between (figure 10.1). These clear lanes are typically one to three miles apart, with the wind somewhat stronger and more veered in them than under the cloud lines.

Even in the absence of cloud, or beneath a uniformly gray and cloudy sky, wind bands will be found over the open sea. The difference in strength between adjacent light and strong bands may be anything from 10 to 25 percent, and their distance apart from one to five miles. Near the coast the position of the bands will normally be fixed by some feature of the coast or the coastline itself, particularly when the wind is nearly parallel to it. Well away from the coast, the bands may move slowly because of the component of the geostrophic wind across them (figure 10.2). Thus, if you are sailing in the open sea, at least five miles from land, and the wind is lighter than it should be, it is advisable to sail on port tack (or jibe) until you find the stronger wind. Having found it, change to starboard tack to stay in the stronger wind as long as possible.

Figure 10.1.

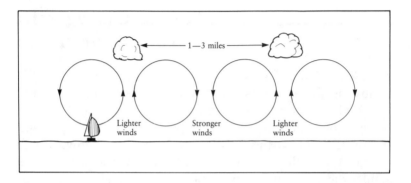

Figure 10.2.

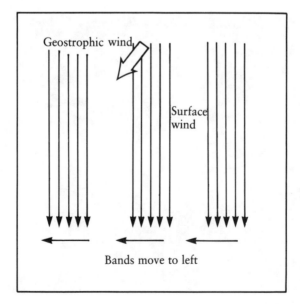

If the geostrophic wind is less than about ten knots the bands tend to deform, and sometimes large (five to ten miles across) eddies appear in the wind. Don't throw the weather forecast overboard in disgust. It cannot include this sort of detail. Look for a return to the original wind within an hour or two.

Changes in water temperature

A sudden change in water temperature of a few degrees is almost as significant as a coastline in influencing the wind. Over colder water the surface air will be cooled and become more stable; there will therefore be less vertical mixing and friction will cause the wind to back and slow down. Over warmer water the surface air will be heated and become less stable, there will be more vertical mixing, and the effect of friction will be more easily overcome, so the wind will be stronger and more veered.

The zone or dividing line between cold and warm water will act as a coastline. There will be a bend as the air moves from one to the other, the bend being always on the side downwind from the transition. For winds blowing along the transition, a zone of convergence and stronger winds—or divergence and lighter winds—will be found, depending on the direction; also, the wind will be generally stronger over the warmer water than over the cold.

If the stability of the air is critical (that is, so close to neutral that it is unstable only over the warmer water), the differences in wind speed between the two areas could be as high as 25 percent, but the areas must be several miles across to be fully effective.

It is common to find water temperature changes of 7° to 9°F, particularly near estuaries or where there are large areas of tidal upwelling.

Probably the best documented evidence of this influence of the temperature difference between the air and the sea on the wind, and of course on the waves as well, relates to the Gulf Stream. The transition from light wind and smooth seas in the cooler inshore waters to relatively strong wind and rough water in the Stream is sometimes dramatic. Measurements made recently revealed an increase on one occasion from 10 to 20 knots accompanying a temperature rise from 61°F to 75°F over a seaward distance of 10 to 15 miles; on another occasion the wind increased from 10 to nearly 30 knots over only 2 or 3 miles as the water temperature rose 11°F.

11

Wind, Tide and Current

Changes in the tidal streams influence the winds in three ways:

• through changes in drag on the wind,
• through changes in water temperature,
• through changes in the temperature of the shore.

Change in drag on the wind

The difference in wave structure resulting from a change in direction of the tidal current can be dramatic. When the tide is running with the wind the sea is smoother than might be expected for the wind experienced, with the waves relatively long and low. Conversely, when the tide opposes the wind the sea is much rougher for a given wind speed, with the waves short and steep. The effect of a steep chop on boat performance may well mask the influence of the change in roughness on the wind itself, whereby there is more wind for a given pressure gradient when it is in the same direction as the tide than when it is opposed.

An extreme example of this sort of thing in a much-used waterway is at the ends of the canal between Buzzards Bay and

Cape Cod Bay. Notorious West Coast examples are found at the entrance to San Francisco Bay and at Admiralty Inlet in Puget Sound. In any of these cases, the short steep waves that develop when wind and tide are opposed can all but kill the boat's speed through the water and require you to hang on with both hands. If you are going with the current the resulting misery is at least short-lived.

Change in water temperature

A change in tide is often accompanied by a change of water temperature, up or down depending on whether the ebb or flood is from a warmer or colder source, or as a consequence of upwelling of colder water from beneath the surface. Colder water leads to colder air near the sea surface, and thus increased stability and a lighter, more backed wind at the sea surface for a given geostrophic wind. Warmer water leads to warmer air near the sea surface, decreased stability, and a stronger, more veered wind for a given pressure gradient. The clear message is: If you want the stronger and more veered wind, sail in the warmer water; if you want the opposite, sail in the colder water.

The tidal influence on water temperature is particularly noticeable where it varies according to the season. River water is normally relatively warm in the summer and cold in winter. On a flood tide the ocean water dominates with its comparatively uniform temperature, but on the ebb the river water spreads out into the estuary. In winter and spring, the river water is colder and will tend to sink, and may not be found at the surface more than a mile or two out to sea. In summer and autumn the river water is warm, and this combined with its low salinity induces it to float above the denser ocean water, perhaps spreading out many miles from the river mouth.

Change in shore temperature

The flooding by cold water of large areas of sun-heated mud-flats or sand changes significantly, and often suddenly, the local sea breeze generating forces. The resulting effects on wind are normally experienced within two to three miles from shore, but can be of considerable importance to competitors in local races and to sailors entering or leaving harbor. These effects can be expected wherever there are extensive tidal flats adjacent to cold ocean water.

Local currents

One other interesting feature of warm river water overlying cold ocean water is its response to the wind. Typically it will overlie the colder water to a depth of a few feet at the most, and will respond independently to any wind blowing over it. Its manifestations are:

• waves characteristic of shallow water, and
• a local wind-driven current independent of any deep ocean current.

It is not unusual to find in summer and autumn local wind-driven currents in warm and low-salinity surface water many miles from its original estuary source. One of the more famous of these is at Acapulco, discovered by one of us prior to the 1968 Olympic yacht races. Here reputably unpredictable and conflicting currents within five miles of the coast were found to consist of a half-knot current in ocean water with a surface temperature about 80°F alongside an opposite 1- to 2-knot current in a shallow layer of low-salinity water with a surface temperature near 90°F, originating from the Verde River some 100 miles down the coast and driven by local breezes.

12

Waves and Swell

It is customary to separate waves into two types: wind waves (often called "sea") produced locally by the wind blowing at the time, and swell waves, which are generated by the wind somewhere else. The "somewhere else" may be thousands of miles away.

The height and distance apart of waves depends on:

• the strength of the wind,
• the length of time it has been blowing,
• the fetch, which is the distance the wind has been blowing over the water,
• and in the case of swell, the distance the waves have traveled.

Wind waves

Locally produced waves are generated very quickly, within an hour or so. Typically a wind of ten knots blowing over a distance of six miles will produce waves of about one foot after about two hours. But height will go on increasing until the limit of about two feet is reached after about six hours with a minimum fetch of 30 miles. Stronger winds require a longer time and a longer fetch before the limiting height is achieved. For example, with a gale of 35 knots, a sea of 20 feet must be expected after 20 hours with a fetch of about 300 miles.

111

The wave front will often lie at an angle of a few degrees to the left of the wind direction, which will tend to make port tack a little faster than starboard. If the wind direction is changing, as with a sea breeze for instance, this angle between the wind and the wave front may be quite large. With the passage of a trough or front there will often be two distinct sets of waves, the angle between them being the difference in wind directions either side of the trough. If the winds are strong on both sides of the trough, the resulting sea can be chaotic, confused, and dangerous.

The time taken for wind waves to decay depends on how long the wind that produced them was blowing. A fair rule of thumb is that the generation and decay times are similar.

If the wind is against the current, the wave length shortens and the waves steepen, while if the current and wind are in the same direction the waves are longer and flatter. A strong current opposing a strong wind, as happens for example when a northerly or northeasterly gale is blowing against the Gulf Stream, produces a very steep and dangerous sea. In this case "the elephants are marching," as the saying goes, and boats crossing the Stream through horrific seas may suffer broken crockery, injury to crew members, or major damage to the vessel. Local examples, as in tidal races around headlands or in confined waterways, tend to be discomforting rather than damaging, but caution is still appropriate. Wind and current in the same direction increase the wave length considerably and make for much easier sailing.

Swell waves

The most important swell waves are those produced over vast areas of open ocean. They contain a great deal of energy and take a long time (days) to decay.

Swell waves always become steeper as they approach a shore, when the depth of the water becomes less than a twentieth of the wave length. This happens for wind waves as well, but except for gale conditions not in the depths where you are likely to be sailing.

It is useful to recognize that the longer swell waves travel faster than the shorter wind waves; for example, they travel ahead and give advance warning of an approaching cyclone. The arrival or absence of swell provides a clear distinction between the advance of a local thunderstorm and an approaching deep low. A threatening sky with increasing black clouds cannot be part of an existing large wind system if there is no swell propagating forward from it, so any strong wind will be temporary. Increasing swell from the direction of advance of the storm clouds suggests an approaching low with a large area of strong winds coming your way.

If there is swell for a long time without significant change the interpretation is uncertain: the low may be approaching but very slowly, or it may have already passed by.

Wave heights

When oceanographers speak of the wave height they normally refer to the average height of the one-third highest waves in a wave record. This is known as the "significant wave" height. Within any wave sample observed at sea there will always be some waves higher than others. Statistical analysis shows us that for a significant wave height of five feet:

• one wave in 200 must be expected over eight feet,
• one wave in 1000 must be expected over nine feet, and
• one wave in 2000 must be expected over ten feet.

So wherever you sail you must expect some waves to be bigger than others, and even the so-called freak wave must be within your expectations. The risk of being pooped or knocked on your beam is always there in gale or storm conditions, so be prepared.

13

Looking for Wind

If you are totally becalmed there is clearly nothing you can do to look for wind. You just have more time to study the weather, so make the most of it. The following are some of the things to think about:

- If you are within a few miles of the coast, will there be a sea breeze (by day) or land breeze (by night)? Or are you very temporarily in the calm zone associated with a developing sea breeze?
- If you are many miles from land, a falling barometer, increasing high cloud, and increasing swell all tell of deteriorating weather and probably wind quite soon—more, perhaps, than you want. If you have been parked near the center of a high, a line of low cloud (probably stratocumulus) moving slowly toward you will herald a gentle breeze from a direction to the left of the direction of approach.
- A change in tide sometimes brings a little wind because it changes the water temperature.

If you have a little wind and want more, there are several possible courses of action:

1. If you are in the open sea and your barometer is reading high, make toward lower pressure on the basis that the stronger winds are normally associated with low rather than high pressure. Applying Buys-Ballot's Law, this means sailing on port tack or jibe until you get the increase

in wind you are looking for. If the high is progressing along at the usual rate, you will not have sailed long before stronger geostrophic winds develop in your area.

2. If you are near the coast, try to work out from your knowledge of coastal effects and of sea and land breezes where the stronger wind will be. For instance, if the wind is blowing along the coast and the land is on your right (back to wind), there will always be more wind (except for the sea breeze calm zone) within about three to five miles of the coast, or within about three miles of a line linking headlands if the coast is wavy. If the land is on your left, stand well off except when a sea breeze is likely.

3. Water temperature is particularly important when the wind is light, and what wind there is will tend to keep to the zones of warmer water. In winter and spring try to avoid the colder coastal waters. Upwelling of water near shore, due to tide or other currents, provides the coldest water, but considerable local experience is necessary to judge where it is.

4. If you are in the open sea and the forecast suggests you should be in a stronger wind, and your own weather maps support this, then maybe you are caught in a light-wind band. Your best chance of finding the band of stronger wind is by sailing on port tack until the wind has increased—that is, sailing toward the direction from whence the band is coming—and then tacking to starboard to stay in the stronger wind as long as possible.

14

Tropical Cyclones

There is an important kind of cyclone that does not depend on horizontal contrasts of temperature between colder and warmer air. This is the tropical cyclone, called a *hurricane* in the Atlantic Ocean, Caribbean Sea, Gulf of Mexico, and eastern Pacific Ocean, if its maximum wind speed is at least 65 knots. The same kind of storm is called a *typhoon* in the western Pacific Ocean. Unlike storms in higher latitudes, which are strongest during winter when temperature contrasts are most pronounced, tropical storms are crucially dependent on high sea-surface temperatures and are therefore seen during summer and early autumn, when the seas are their warmest. In fact, during the rest of the year they rarely if ever occur in the western hemisphere. In the western Pacific, however, typhoons occur in all months, although they are relatively rare during the colder season.

Necessary ingredients

a) The sea-surface temperature must be at least 82°F. At these high temperatures the amount of moisture in saturated air is extremely sensitive to temperature. For instance, over a sea-surface temperature only 5°F cooler the water-vapor content at saturation would be nearly 20 percent less. A rich high-temperature mixture is the essential fuel for a tropical cyclone. Without it, the storm quickly dies.

b) The developing storm must be no closer than about 400 nautical miles to the equator, or no farther south than about latitude 7° N. The cyclonic circulation spins up in the same way that it does in an extratropical cyclone—that is, by the rightward Coriolis deflection of an air current converging toward a central point of low pressure. The Coriolis force does not act at all on the equator and is too weak to be effective less than 400 miles away.

c) The air column must be conditionally unstable or nearly neutral through the entire depth of the atmosphere up to around 50,000 feet. Otherwise, cumulus clouds, which must act to convey heat and moisture up to great heights in the developing storm, are suppressed.

d) The pattern of air flow near the surface must contain some weak initial circulation with a horizontal dimension of a few hundred miles to act as a triggering mechanism. A random thunderstorm will disturb the surface air flow, but not on a large enough scale.

e) The vertical shear of the wind, from near the surface to about 40,000 feet elevation, must be relatively weak—less than about 30 knots in magnitude. At latitude 15° N a thermal wind of 30 knots over this depth would require a horizontal temperature contrast of no more than 1°F in 200 miles, so you can see how nearly horizontally homogeneous the air must be. A greater shear tends to disrupt the organization of the storm circulation, which is roughly symmetrical about the center.

Where and when

Maps of monthly-average sea-surface isotherms show when and where, north of latitude 7° N, tropical storm development is something to consider. In the western hemisphere, the possible regions in September (figure 14.1) are the central Atlantic up to about latitude 15° N, the western Atlantic up to about 30° N, and all of the Caribbean Sea and the Gulf of Mexico, as well as the extreme eastern Pacific between approximate latitudes 10° and 20° N. The vertical stability of the atmosphere does not prevent deep cumulus convective clouds in any of these regions during the summer and early autumn. In the northeasterly trade wind belts during the rest of the year, a pronounced stable layer puts a cap on cumulus clouds at a height of no more than 10,000 feet, thus inhibiting the deep transport of heat and moisture required for a tropical storm. Relatively strong vertical wind shear tends to prevent initial development of cyclones in the central and eastern Caribbean, even though these regions are often traversed by storms that start farther to the east.

Three kinds of initial disturbance are large enough to serve as the trigger for a tropical storm. The first of these starts as a

Figure 14.1. *Average sea-surface isotherms in September. The shaded areas show where the vertical wind shear from the surface to 40,000 feet is greater than 30 knots.*

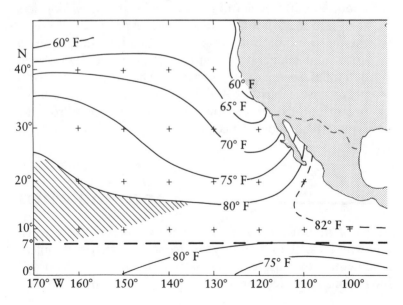

wave in the easterly winds that peak at a height of 10,000 feet over central Africa, just south of the Sahara, during the summer. Since the air up to this elevation is much hotter over the desert than over the tropical coastal regions to the south, the north-south temperature gradient is "upside down." The warmer air is to the north and so the thermal wind here is from the *east* rather than the west (see Chapter 3). Troughs and ridges develop and move westward, accompanied by clouds, showers, and thunderstorms downstream from the troughs. This is consistent with the pattern of bad weather associated with upper troughs in higher latitudes (see Chapter 5), except that the bad weather is *west* of the troughs rather than east. These waves are common from July through September, crossing the African coast as frequently as one every other day.

As these African waves move out into the Atlantic they lose the temperature contrast that enabled them to form, and so begin to weaken. Most die away without producing more than scattered showers and thunderstorms (figure 14.2). A few,

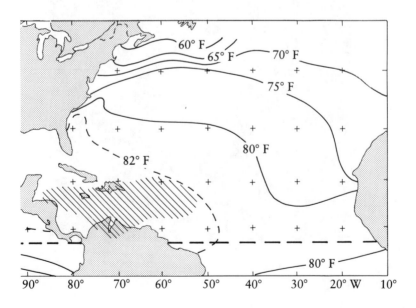

Figure 14.2. *An easterly wave dissipating over the central Atlantic Ocean.*

Figure 14.3. *A well-organized easterly wave east of the Caribbean Sea.*

Figure 14.4. *A tropical storm brushing the northeastern Caribbean Sea.*

Figure 14.5. *This mature hurricane found its way through the Yucatan Channel.*

however, make it across the ocean and appear in the western Atlantic or Caribbean as *easterly waves*, now in an environment of weak *westerly* thermal wind, accompanied by an organized zone of cloud and rain with embedded showers and thundersqualls. This bad-weather zone is here found *east* of the wave trough (figure 14.3), again consistent with the general principle that the ascending air is found downstream in the thermal wind from the trough or low center. A small fraction of easterly waves amplify in the layer of air near the surface, forming a closed vortex that then strengthens from its initial state of *tropical depression* to the status of *tropical storm* (figure 14.4) if the maximum winds exceed 35 knots and perhaps to hurricane (figure 14.5) should the winds reach the 65-knot threshhold. It is customary to give names to depressions that achieve at least tropical-storm strength, in alphabetical order from the start of the season.

Occasionally an easterly wave is thought to continue its travel across the Caribbean and then to cross Central America, subsequently producing a tropical cyclone in the eastern Pacific. More often, however, storms in this region form from disturbances within a low-pressure zone of convergence in the surface wind, which extends westward from the coast of Central America approximately along latitude 12° N. This east-west trough separates the northeasterly trade winds of the northern hemisphere from the southeast trades of the southern hemisphere, which have become light southwesterlies after their northward crossing of the equator. This trough is the frequent site of thunderstorm systems and other disturbances large enough to initiate tropical cyclones.

Near the beginning and near the end of the tropical-cyclone season, an upper trough from middle latitudes, sometimes accompanied by a weakening cold front at the surface, may settle into the southwestern Atlantic, the western Caribbean, or the Gulf of Mexico and become stationary for a number of days. Occasionally one of these systems becomes transformed into a tropical cyclone, which is then indistinguishable from one whose origin is strictly tropical.

Despite the number of circumstances through which tropical cyclones can form, storms of appreciable strength are quite rare: In the region comprising the Atlantic, the Gulf, and the Caribbean, only about ten systems annually reach tropical-storm strength, of which, on average, six become full-fledged hurricanes. The greatest likelihood of a storm here is in August and September. In the eastern Pacific the relatively limited eligible area is quite productive: fifteen tropical storms per year, of which eight strengthen further to hurricanes, with maximum frequency about a month earlier, in July and August.

How the storm works

Suppose we take a vertical slice through the center of a hurricane like the one shown in figure 14.5. The main circulation of the storm winds is rotary, coming out of the left portion of the cross section shown in figure 14.6 and into the right portion. The highest speeds (up to 150 knots) are reached near the surface in a ring-like *eye wall* of torrential rain and towering clouds surrounding a calm and nearly clear *eye*, which may be very narrow and in which sea-level pressure may fall to extraordinary depths.

Figure 14.6. *A vertical cross section through the center of a hurricane.*

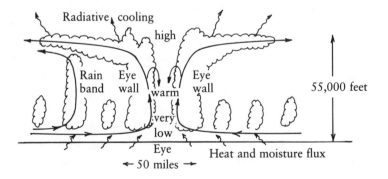

Radiative cooling
high
Rain band Eye wall warm Eye wall
very low
Eye
55,000 feet
Heat and moisture flux
← 50 miles →

The physical processes that sustain this powerful whirl are tied to the relatively modest radial and vertical circulation seen in figure 14.6, typically 20 knots or so. This circulation has four parts. The inflow is concentrated in the lowest 3000 feet. Here the fuel for the storm is received from the sea. As the air flows in toward lower pressure, two things happen to encourage the vigorous transfer of heat and moisture from the sea surface into the atmosphere. First the temperature of the air tends to cool adiabatically, but heat transfer from the warm water maintains the air temperature despite the often large fall of pressure. The rotary part of the wind gets stronger as the Coriolis force turns the inflow to the right more than enough to balance the loss of wind energy to surface friction. Most important, white-caps and blowing spray develop to hasten the evaporation of water into the air and raise the humidity even above what it is in the sultry air on the periphery of the storm.

Next, the vigorous ascent in the cumulus clouds that compose the eye wall causes a massive release of latent heat energy, most of which is transported upward to near 55,000 feet. Above the surface layers the central portion of the storm is warmer than its horizontal surroundings, although the temperature profile shows a decrease with height close to the moist-adiabatic lapse rate. Thus, whereas the surface inflow is at a temperature between 75°F and 80°F, about the same as in the surroundings, the rising cloudy air has cooled to about −60°F by the time it reaches the 40,000-foot level. But this is still 5° to 10°F less cold than the surrounding air at this level, which has not received the heating due to water-vapor condensation. A small amount of the main updraft branches off to descend into the eye itself, which becomes much warmer than the air outside. The warm core of the storm is so pronounced that it has become a high at elevations near 50,000 feet, except for a tiny region in the eye.

At the top of the updraft the air diverges and acquires *anticylonic* rotation in a great round shield (figure 14.5) of cirrus cloud, due again to deflection of the outward flow by the Coriolis force and consistent with the generally high pressure in the storm region at these elevations. A series of closely spaced satellite pictures often shows the counter-rotation of the incoming lower and outgoing higher clouds, cyclonic below and anticyclonic above. The heat energy received from the sea and released largely during condensation of water vapor in the eye wall is now slowly lost by radiation to space from the cloud tops in the outflow layer.

The circuit is completed by a broad and gentle descent (not shown in figure 14.6), which may occur at a great distance from the storm. The entire system runs like a great heat engine, with the energy input at the high temperature of the sea surface and with the heat loss at the cold temperatures of the high cirrus cloud near the tropopause. As in mid-latitude storms, the kinetic energy of the winds comes from a tendency to lower the center of gravity of the air, brought about by an average ascent of warmer air and descent of colder air. But in the tropical storm the warmer air is in the middle rather than on the southern flank, and in the tropical storm the connection to the sources of thermal energy is particularly close and direct.

The behavior of the hurricane is not actually in such a steady state as this account implies. In particular, an outer rain band will typically begin to develop (see figure 14.6), accompanied by a secondary updraft and a local maximum in the rotary wind. This band sometimes wraps around the entire storm and shrinks in toward the existing eye wall, with an increase in its own rotary wind maximum. The old eye wall then dies away and is replaced by the new one, which contracts and may be replaced in turn by a yet newer cloud ring. The cloud structure and the maximum winds in the overall storm, as well as the lowest central pressure, may fluctuate strongly over a day or sometimes over several hours as these evolutions occur.

The track of the storm

The precise path that the tropical cyclone takes is of particular importance not only because of the life-threatening wind and waves that it can produce but because the central region of truly dangerous conditions is so small. The rules are quite different from, and fundamentally simpler than, the rules for mid-latitude storms. In that case the cyclone moves according to the orientation of areas of convergence and divergence in the flow pattern near the surface, which are associated with areas of warm and cold advection, and only coincidentally appears to be more or less "steered" by the flow aloft. The tropical cyclone has thermal symmetry about the center and occurs in an environment with little overall horizontal temperature gradient, so these effects are of little consequence. The tropical cyclone, on the other hand, is clearly embedded within an overall flow pattern of scale larger than that of the storm, and it is quite literally steered by that flow, except for a westward drift, usually small, due to the earth's rotation.

Since these storms, except for the ones that develop in old troughs from mid-latitudes, develop initially within or at the edge of the easterly trade wind belts of the subtropical or tropical latitudes, their initial motion is toward the west at 10 to 15 knots, as shown in figure 14.7. When they approach the longitude of the western extremity of one of the great subtropical anticyclones (such as the Azores High in the North Atlantic), the motion veers slightly toward northwest. Storms in the Atlantic-Caribbean-Gulf region may make landfall between Central America and southeastern North America on

Figure 14.7. *A sample of hurricane tracks.*

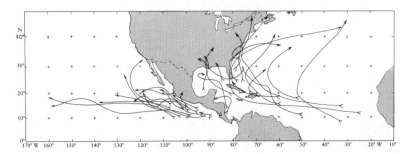

one of these headings, with subsequent quick weakening or complete dissipation. In the eastern Pacific, such storms usually weaken because of an encounter with cooler water before reaching the Hawaiian Islands.

Some of these cyclones (figure 14.7) follow the large-scale flow as it turns northward and then northeastward on approaching mid-latitudes. Those in the Pacific thus *recurve* into the west coast of Mexico, with effects reaching as far north as southern California on rare occasion. Those that similarly follow the flow around the western end of the Atlantic subtropical anticyclone may remain over the Gulf Stream and threaten the entire east coast of North America up through the Canadian Maritime Provinces. If the storm is moving fast enough it may cross the cooler waters of the continental shelf without weakening much before striking shore.

If the recurvature of an Atlantic storm is sharp enough it will spare the continent and travel out into the central ocean. In this case, or sometimes in the case of large-scale southwesterly flow in advance of an pronounced upper trough, the cyclone finds itself in an environment with strong horizontal temperature contrasts and accelerates along its track. In this case it is likely to become transformed into a cyclone typical of mid-latitudes, sometimes a very severe and fast-moving one which must be treated with the respect due any hurricane. Speeds of advance of such a storm can exceed 50 knots, adding to the hazard of dealing with it. If it follows this course it will eventually weaken and lose its identity in the manner of mid-latitude storms.

Sometimes in the latitudes near 30° N the large-scale steering currents become weak and variable. In this event the path of the storm is erratic (figure 14.7), containing sharp turns, loops, and reversals as the steering flow evolves in time, before finally adopting a definite course. Hurricane Betsy brushed the Bahamian island of Eleuthera in September 1965 but then stopped and looped just after it apeared to have passed, returning to strike with full force three days later on its way to the Florida Keys and the Gulf of Mexico. You can find Betsy's errant path in figure 14.7.

The best way to deal with tropical cyclones is to stay out of the susceptible parts of the oceans at the dangerous months of the year. Ways of maneuvering should you find yourself in the path of one are set out in the next chapter.

15

Weather Hazards

What is that black cloud to windward? Does it indicate an approaching gale, a thunderstorm, a cold-front squall, dense fog, or is it just a patch of dirty, smoke-laden air? There are a number of questions you can ask that will at least help toward an answer and may rule out some possible hazards:

• What was the latest weather forecast?
• What is the barometer doing?
• Have the direction and speed of the wind changed much in the last hour or two?
• Have the waves changed? Is there any swell?
• What have the higher clouds been doing? Have they been typical of those ahead of an approaching cyclone? Ahead of a thunderstorm? Did you notice which way they were moving?

All this information you should have in your ship's log. It is important to refer to the *facts*, not casual recollections or guesses as to what may have happened recently.

Gales and more

Gales due to cyclones last from 12 to 24 hours or more and do not spring up without warning. Forecasts of gales show a good amount of reliability, although they may lack precision as to time and location. With the information in your log, which will enable you to answer the above questions, you should never be caught unprepared. Pay particular attention to:

1. *The weather forecast.* If the forecast has mentioned the possibility of a subsequent warning for a gale (34 to 48 knots) or storm (48 knots or more), keep a listening watch for the high-seas forecast or for one of the continuous broadcasts of NOAA Weather Radio or Weatheradio Canada.

    Most mid-latitude gales during the summer are not particularly severe. They are likely to produce rough, uncomfortable conditions but are seldom a threat to life if the boat is handled with prudence and common sense. Storms during the colder season, in the western Atlantic from November through March (and in the rare summer event like the storm in the 1979 Fastnet Race), on the other hand, can be extremely severe and are to be avoided by small craft, even if that means staying close enough to shore to be able to reach reliable access to shelter within 12 hours. Remember that the inlet you have in mind may be barred by dangerous seas by the time you reach it. If a hurricane warning is in the picture, you should not be out there. If you are, for whatever reason, do your best to avoid the small inner core of truly destructive winds and seas. Survival is the issue here.

2. *Your barometer.* There is a twice-per-day tide in the atmospheric pressure, just as there is in the sea level. In the atmosphere it is always low at 0400 and 1600 local time, and high at 1000 and 2200 local time. A change of 1 millibar (.03 inch) in the three hours prior to one of these times is only normal; the rise to the morning high and the fall to the afternoon low are the most marked. Changes much larger than this, however, mean that you are likely under the influence of a cyclone and that a gale is possible. Numerous southwesterly or southeasterly gales occur with rapidly falling pressure in advance of a low, and when the barometer reaches bottom the hazard has not passed.

Many of the most severe gales are northwesterly or northerly ones, with rapidly rising pressure after the trough has gone by.

3. *The wind.* If the light westerly winds of a ridge of high pressure give way to a strengthening southerly or southeasterly, the gale may be upon you in a few hours. Sometimes the southerly remains only moderate, but freshens to gale force upon the shift to northwest with the passage of a cold front or trough 12 hours or so later. Then you have at least the comfort of anticipating that the worst is over and that you may be enjoying an anticyclone tomorrow.

When a freshening wind associated with a mid-latitude cyclone is from the direction toward which you are sailing, go on starboard tack. That is, if the wind is southwesterly, sail south; if northwesterly, sail west. In either case, provided the cyclone follows the usual track toward the northeast, you will be out of it sooner than if you had made the other choice. Be as certain as you can that the low is proceeding normally, however, and not just drifting uncharacteristically westward as part of a deep, mature system. One of us made the conventional choice but paid the price on a passage from Bermuda northward to New England. Confronted with a norther he tacked west, only to move parallel to a deep cyclone that then transformed into a tropical storm, in a reluctant race toward Chesapeake Bay. It was one of the rare occasions when the forecast was not helpful, and it is difficult to say how events might have been anticipated.

If forced to deal with a tropical cyclone, starboard tack is a good general rule, since it will tend to carry you out through the counterclockwise spiral of surrounding winds, away from the severe conditions near the center. If the storm is coming directly at you and there is no possibility of outrunning it (which will be the case unless it is moving extremely slowly), then head for the side to the left of its track—that is, the south side if it is moving westward in the tropics, the west side if it is moving northward in the process of recurving, and the north side if it is moving eastward in mid-latitudes. The reason for this tactic is that the winds of the tropical cyclone can be viewed approximately as the sum of a counterclockwise vortex and a broad basic current which determines the direction of the track of the storm. You can see that to the right of the track the vortex wind and the basic current reinforce each other, whereas to the left they oppose each other. Hence the less strong winds are to be found to the left of the track direction.

This rule cannot be used with a mid-latitude cyclone, because its track is determined by the placement of regions of cold and warm advection. It coincidentally appears to be steered by the flow aloft, but any broad surface air current in which it may be embedded will be unrelated to the direction of its track. The strongest winds may be in any quadrant of the storm.

4. *Waves.* Those produced by the local wind tell you nothing more than the wind itself, but swell, produced by distant winds, can be a useful predictor. Any waves larger than what seems appropriate for the local wind, in fact, deserve attention, but first try to satisfy yourself that they are not the result of a local current running against the wind. They may indicate the direction from which the stronger wind that made them will blow. On the West Coast, a prominent westerly swell sometimes indicates the approach of a broad region of strong westerly winds, which may last for a day or more, rather than a few hours. On the East Coast, an easterly swell is often produced by a storm which has passed by, leaving only this weakening reminder of its power. If the swell has a period of more than about five seconds, it may have been produced by a gale or hurricane a great distance away which represents little threat to you. Even so, another check on the forecast is worthwhile.

5. *Clouds.* A solid shield of high clouds rising in the sky from the direction of the cyclone means that the weather system is likely moving toward you. If the direction of movement of the upper clouds is rotated clockwise from the direction of the surface wind or low clouds, then an increase of wind is indicated. In this case you can see that the lower-level

wind is blowing across the thermal wind from warmer toward colder air: there is warm advection and the cyclone is moving toward you. If the high clouds are moving rapidly, then the thermal wind is strong, and so is the contrast between warm and cold air, and so therefore, in all probability, is the oncoming storm.

Thunderstorms

If the dark mass of threatening cloud occupies only a part of the sky in its direction, a thunderstorm is approaching, with stormy conditions lasting only an hour or so, a fraction of the duration of a gale. Nevertheless, while it is in progress, the severe thunderstorm can produce winds of hurricane force, large hail, and torrential rain in addition to the lightning strokes. It is not usually possible to forecast exactly when and where a thunderstorm will strike, but the general conditions are readily anticipated, and where radar surveillance is particularly good specific forecasts can sometimes be obtained on NOAA Weather Radio for about two hours in advance.

Thunderstorms occur over the sea much less often than they do over land, by a factor of about ten overall. The surface layers of air over water do not receive the strong charge of heat during the daytime which results in the strong instability responsible for frequent extremely deep and vigorous convective cloud development over land. Dangerous thunderstorms do occur, however, over the sea.

There are important differences from place to place over water, as shown in figure 15.1. The highest thunderstorm frequencies occur over the warmest water, which maintains instability in the overlying atmosphere. These favored areas are the easternmost Pacific south of latitude 20° N, the Caribbean, the Gulf of Mexico, and the region along the Gulf Stream nearly up to New England. Notice that thunderstorms hardly ever occur over the cold water offshore along the Pacific Coast of the United States and Canada; here the frequencies amount to about one day in three years. The incidence along the coastal regions of Central America and the east coast of the United States is high in part because of the high frequency of storms over the nearby land, which may drift out over the adjacent water or redevelop there. Even those that break out over coastal waters received their unstable charge over land. Relatively high frequencies due to this reason are maintained even over the cold water of the Canadian Maritimes.

There are seasonal differences, too. In most of the areas of high frequency the maximum is in summer, as it is over the adjacent land. In the Caribbean Sea and eastward, however, thunderstorms are most frequent in the fall, when the water is still very warm and easterly waves and tropical cyclones are at their most vigorous. In mid-latitudes in the central Atlantic and the eastern Pacific, the occasional thunderstorm may occur in winter when a very cold air mass originally from a continental source has been sufficiently heated and moistened by the sea.

The time of day makes a difference, but only along the immediate shore does the high frequency occur in the late afternoon or early evening, as in most places over land. Farther offshore, more often than not, the storms go up when the sun goes down and subside shortly after dawn. Coastal radar observations show this effect within their reliable range of 125 nautical miles from the Gulf Coast around Florida and up to New England, and the nocturnal maximum is observed farther at sea as well.

Figure 15.1. *Annual number of days with thunderstorms.*

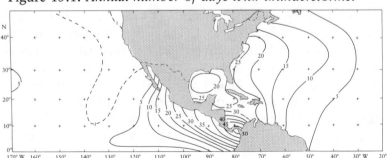

Many thunderstorms are very short-lived, lasting no longer than an hour. The updraft builds the typical cumulonimbus cloud, and the subsequent rain shaft brings a downrush of cool air with it, quenching the storm. The spreading of the cool air as it approaches the surface can produce a brief squall, as mentioned in Chapter 6. Such storms are widely separated from each other and can usually be avoided altogether.

These storms or even lines of innocuous-looking cumulus built to 20,000 feet or so can produce waterspouts. These weaker relatives of the deadly tornado occur most frequently over very warm water, rarely last longer than a few minutes, and often appear in succession from the line of cloud over a period of an hour or two. They are fascinating to observe, but

Thunderstorm development along the mid-Atlantic and south-eastern coasts. Note the new storms growing off the Florida coast, in interior South Carolina, and around Chesapeake Bay, while a large, mature system patrols the Gulf Stream east of the Carolinas.

it is just as well to keep your distance, since the occasional one can produce damage should you happen to cross its path.

The thunderstorm *system* presents much more of a threat. If the direction and speed of the wind vary in the right way upward through the depth of the cloud layer, the rainy downdraft occurs adjacent to the updraft rather than in it so that the system is self-perpetuating and can continue through a succession of updraft pulses for 12 hours or more. A number of thunderstorm cells of this type often line up over a distance of 200 miles or more. Whether in the tropics or in mid-latitudes, these systems have a similar structure, seen in a slice (figure 15.2) through an active storm in the line. In the tropics, west is to the left and east is to the right. In mid-latitudes southeast is to the left and northwest is to the right. The system is moving from right to left in both cases.

As the storm approaches you, the wind is blowing, often only gently, toward the cloud mass. If there appears to be a weak spot in the line of cloud, nothing is to be gained by heading for it, because this gap will probably be filled in by a vigorous new cumulonimbus cell by the time the storm reaches you. A thin veil of cirrus cloud, the leading portion of the anvil, comes overhead while lightning is visible in the darker cumulo-

Figure 15.2. *Vertical cross section through a cumulonimbus cloud in a line of thunderstorms.*

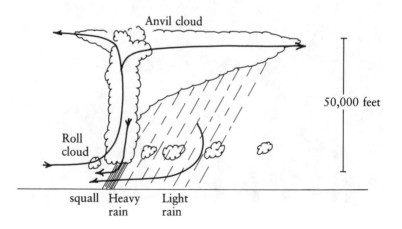

nimbus in the background, reminding you to consider whether your boat's grounding, providing a safe metal path for electric charge from the masthead to the water, is adequate. If in doubt, you may wish to trail a length of chain overboard from a shroud. Often a band of low, smooth-looking cloud can be seen just at the leading edge. This is the roll cloud. Lowering and securing the headsail and tucking a reef in the main at this time is a good idea, because it is difficult to say how hard the squall will be. The faster the roll cloud advances toward you, the stronger the accompanying blast is likely to be. The direction of the squall wind will be from the direction of the heaviest rain. It may last only ten minutes—even that will seem like an hour—but it may be reinforced by renewed bursts of downdraft and heavy rain over a distance of as much as ten miles. On rare occasion one of these downbursts may be extremely violent over a very small area. This phenomenon may have been responsible for two tragic sinkings of sailing ships in the past few years. Eventually, the rain will lighten, the wind will ease, and the skies will clear. The light wind you had prior to the storm may even return.

When conditions favor the development in the next few hours of severe thunderstorms accompanied by dangerous winds, large hail, or funnel clouds or tornadoes, NOAA Weather Radio will broadcast a severe-thunderstorm (or tornado) watch, alerting you to the possibility. When a particular storm is known or presumed to be accompanied by these effects, a warning will be broadcast for a restricted area over the next hour or so. If you are in range of one of these stations, you should check for the existence of a watch or a warning as soon as you see signs of an approaching thunderstorm.

Cold-front squalls

Some squalls have no direct connection with thunderstorms, although a rumble of thunder may be heard, embedded in a brief period of heavy rain and particularly strong wind. These are found along or ahead of cold fronts, when a low center is passing eastward well to the north of your position.

In one instance, on a summer passage to Bermuda, a steady fresh south-southwesterly had already been blowing for some time (figure 15.3) and skies were cloudy and hazy. About 12 hours before a cold-front wind shift to northerly, a mass of

Figure 15.3. *Half-hourly observations of wind during cold-front squalls in the Western Atlantic. The arrow shows the wind direction, and the length is proportional to speed. These observations were taken in the frontal cloud bands shown in the satellite photographs, along a line between southern New England and Bermuda.*
15.3a. *Cold front encountered during passage to Bermuda. Ship's location shown on photo.*

Ship's time

0800	1200	1600	2000	0000	0400	0800

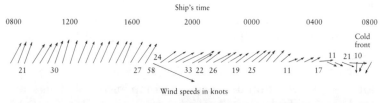

Wind speeds in knots

black clouds brought a violent squall up to 58 knots from the northwest. Thirty minutes later there was a brief lull to 24 knots as the wind swung back, but only to west-southwest. Then the breeze died gradually through a series of surges and lulls while veering around toward west and finally north as the front passed. On the return passage to New England a week later, another cold front was encountered (figure 15.3). This time the south-southwesterly built gradually to the edge of gale force. Then the wind veered perhaps 15 degrees, as it had earlier, but this time it peaked at 62 knots. A half-hour later it had returned to 34 knots, but relief was on the way: within five

15.3b. *Another cold-front encounter, one week after the first. Ship's location shown on photo.*

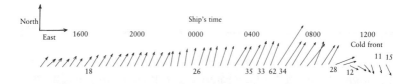

hours the wind was under 20 knots with a veering cold-frontal shift to northwesterly.

In the first case the initial wind shift might have been mistaken for the cold front, but instead it marked only a period of squally conditions ending with a mild frontal shift. In the second case, the southwesterly squall occurred closer to the cold front and represented the culmination of a prefrontal southwesterly blast, with the front itself again benign. This type of structure occurs more often than not during summer in the western Atlantic, but on other occasions a brief period of northerly squalls will occur just after the frontal wind shift rather than clearly in advance of it. In either case, unless the forecast calls for a prolonged gale, the strong winds will likely have eased substantially in an hour. The best advice is to make sure that you are not over-canvased and to wait it out.

Fog

Along the California coast, off the New England coast, and through much of the Canadian Maritimes, sailing in fog is a way of life during the summer; good visibility seems a blessing.

The ingredients for sea fog are simply cold water and moist air. In particular, the dew-point temperature of the air must be higher than the sea-surface temperature. When these conditions are met, the air in contact with the sea is cooled until the water vapor condenses and a fogbank forms. This cooling at the surface tends to produce a temperature inversion in the overlying atmosphere. Thus, the air layer becomes very stable, vertical mixing is inhibited, and the fogbank persists. Only the arrival of a new, drier air stream will disperse it more than temporarily.

In the western Atlantic the weak but persistent Labrador Current brings cold water from the ice-laden northwestern corner of the Atlantic down the east coast of North America (figure 14.1) nearly to Cape Hatteras. A day or two of south-westerly flow following the passage of an anticyclone will bring in air of increasingly high dew point, with values rising to or above the sea temperature. The farther northeast you look the greater the excess of dew point above sea temperature and the more widespread the fog, culminating in the great prevalence of fog along the Grand Banks east of Newfoundland. Certain indentations in the coast enjoy an exemption from this cold because of solar heating of shallow water or discharge of warm water from rivers. These areas include much of the Gulf of St Lawrence, the sounds along the south coast of New England, and the Delaware and Chesapeake bays. In winter and spring, by way of contrast, these water bodies are cold and thus foggy on those occasions when moist air flows northward in advance of a cyclone. The discharge of cold river water into the northern fringe of the Gulf of Mexico maintains a low sea temperature there during the winter, with a substantial likelihood of fog.

Fog and cirrus over the Canadian Maritimes. They are often hard to distinguish in this kind of satellite photograph. Note the signs of cumulus buildup in the clouds east of Delaware and Chesapeake Bays (arrows).

The prevalence of summer fog along the west coast of North America seems a paradox. We can accept that the water is cold, but why should winds from the northwest be so warm and moist as to cause fog to form? The reason is the persistent Eastern Pacific High. The air in the northwesterly winds on its eastern flank—the West Coast—has a tropical origin, starting as southerlies on its western side, subsequently turning to westerlies and then northwesterlies as they move around the high center. Fog so formed over a very long sea track tends to be deeper and more persistent than the Grand Banks type of fog. A nice example appears in the satellite photograph in Chapter 3. Incidentally, the reason for the West Coast's cold water is interesting. Its origin is deep down. A surface wind current is generated by the persistent northwesterly winds, and this current is deflected to the right, away from the coast, by the Coriolis force. The surface water thus driven away is replaced by rising cold water from a considerable depth.

Close to any coast, the behavior of fog can be tricky. Where the coast is indented by bays and peninsulas or is protected by

Particularly dense sea fog, called "haar," impinges on the north coast of Scotland.

offshore islands, a useful lifting or local clearing of a marine fogbank can almost always be found in the lee. Additionally, at night a land breeze may displace a sea fog to a modest distance offshore. A feeble northwesterly at the leading edge of a weak anticylone also can clear fog to a few miles offshore. In these cases, beware the sea breeze. It may simply bring the fog back to shore during the afternoon after a clear morning.

Another nocturnal effect to be aware of is fog drifting seaward from the land on a clear night. If the night cooling over land causes the air temperature to fall below its dew point—the dew point has to be high enough to make this possible—shallow fog will form and drift out on the land breeze over otherwise fog-free coastal waters. The warmer water will cause it to lift and dissipate, and the warmer the water in contrast to the temperature of the land breeze the more quickly the fog will go.

When there is an offshore wind with high land close to the coast, coastal waters are typically clear owing to the dry, adiabatically warmed descending air current, even though a fogbank may be present out at sea. In southern California, a small cyclonic "Catalina eddy" sometimes forms near that island, in the lee of the coastal mountains, in these circumstances . On occasion the southeasterly breeze on the east side of this eddy can bring a fogbank from Baja California up along the coast into southern California.

The best advice when there is the least threat of fog is to establish your position with care before your visibility deteriorates. It is often easy to forget this, because it is such a nice day before the fog arrives, and it is too late afterward if you have to rely on visual reference. There is always a risk of running aground, but the greater threat is being struck by a larger ship. Have a good radar reflector as high as possible in the rigging, and sound the foghorn conscientiously. Stay out of shipping channels if possible, because even though a larger ship may have radar, a small sailboat is a difficult target to separate from the background, and the sight of a "wall of black steel" emerging from the fog at close range is an experience we all wish to forgo.

16

Observing the Weather

Your own log of the weather is a useful adjunct to the weather forecast. The first step in using a weather forecast is to build the best picture you can of what the weather is *now* over the area around you. Your own observations can help you interpret the forecast in terms of the changes that will occur at your position or along your route. You should observe and log the following:

1. *Wind direction* and *wind speed*. These can be recorded as relative to the ship from either the masthead anemometer or a hand-held instrument on deck. Then combine with heading and speed through the water to obtain the true wind. In the absence of an instrument for recording wind speed, a judgment based on the state of the sea is advised.
2. *Atmospheric pressure*. It is the changes that matter most, but it does help to have your barometer reading correctly so that your observation fits the others on your weather map. If you have a barograph, record whether the pressure is rising, falling, or steady.
3. *Visibility*, in general terms (good, very good, poor).
4. *Clouds*. Keep your eye on the sky. Get used to looking at the clouds, the high ones as well as the low ones, and judging their speed of movement—not in absolute terms but merely whether they are moving fast or slowly and from which direction. The clouds hold many clues to the weather in the next few hours.

0800   Wind 260°/18. Pressure 1011 mb, rising. Visibility good, rain showers in distance. Overcast, altocumulus and altostratus. Seas 1-2 feet, swell 15 feet from 190°, period 8 sec. Air 68°F, Sea 75°F.

1100   Light rain. Wind backing. Pressure 1011 mb, steady.

1400   Wind 210°/23. Pressure 1010 mb, falling slowly. Visibility good, rain ended. Breaks in overcast altocumulus and cirrostratus, with scattered stratocumulus beneath. Sea 3 feet, swell 15 feet from 190°, period 8 sec. Air 70°F, Sea 74°F.

1630   Wind shift to 340°/35, gusts to 50. Visibility fair in heavy rain shower. Overcast stratocumulus. Pressure rising rapidly.

1700   Wind easing. Rain ended. Pressure 1013 mb, rising rapidly.

2000   Wind 340°/20. Pressure 1017 mb, rising. Visibility good. Breaks in overcast stratocumulus. Waves 15 feet, southwesterly swell. Air 65°F, Sea 76°F.

Extract from a ship's log.

---

5. *State of the sea*. Note particularly the height, the period (time between wave crests), and the direction of any swell.

When you are making a passage, complete your weather log at least once per watch, and every hour or less if the weather is changing quickly. If you have a thermometer or hygrometer, log their readings as well.

An extract from a yacht's weather log, made during unusually bracing conditions, is shown here.

17

Using Weather Information

Coastal sailors have ready access to a substantial amount of weather information, which when put together enables the best judgment to be made on everything from tactics for winning a race to the most suitable route for a pleasant day's sail. Weather charts in the newspapers and on television all contribute to the background of information necessary to put your own observations and deductions in the context of the major highs, lows, troughs, and ridges. It is worth studying the sequence of weather maps for several days before you set sail, so as to get tuned in to what the weather is doing. Then when you are at sea you will have both a good idea of how the weather systems are likely to evolve and move, and also a good idea of the shape of the isobars, so you can fit the observations you receive into a coherent pattern. It is worth noting that a particular weather mood may persist for a couple of weeks or more, with similar sequences tending to repeat themselves. This just underlines the value of getting tuned in. This tendency toward weather regimes is more prominent on the West Coast than in the Northeast, where the saying, "If you don't like the weather, wait a minute," is not a total exaggeration. Even there, though, it is important to recognize the current tune.

146

Sources of weather information

United States—Coastal and adjacent offshore waters. The National Oceanic and Atmospheric Administration (NOAA) Weather Radio (NWR) provides continuous broadcasts (24 hours a day) of the latest weather information and forecasts directly from National Weather Service offices. Frequencies of 162.550, 162.400, and 162.475 MHz (VHF channels WX1, WX2, and WX3) are used. The locations and frequencies of the NWR stations are shown in figures 17.1–17.4, to which it should be added that there are two NWR stations in Puerto Rico.

Broadcasts appropriate to special requirements for boating, fishing, and other marine activities cover virtually all coastal and offshore areas of the United States as well as the Great Lakes. Each broadcast is tailored to the receiving area within about 40 miles from the antenna, and refers to a variety of interests in addition to the marine ones. The taped weather messages are repeated every four to six minutes and are revised every one to three hours, or more frequently as necessary.

The marine information varies from station to station and from time to time. It typically comprises any pertinent warnings, the latest observations from coastal stations and buoys (sometimes including weather radar observations), a synopsis or description of the position and movement of weather systems from the current weather chart, and marine forecasts for the next 36 to 48 hours. In each broadcast, forecasts are made for the local coastal waters and for one or more of the adjacent offshore areas shown in figures 17.5–17.7.

You should not hesitate to use forecasts and observations directed at aviation interests. These are produced by the National Weather Service and transmitted by the Federal Aviation Administration as the continuous Transcribed Weather Broadcast (TWEB). Figure 17.8 shows the location of TWEB stations likely to be heard from coastal waters of the United States. Frequencies are from 200 to 400 kHz, in the long-wave band used by marine radio-direction finding stations. Of particular

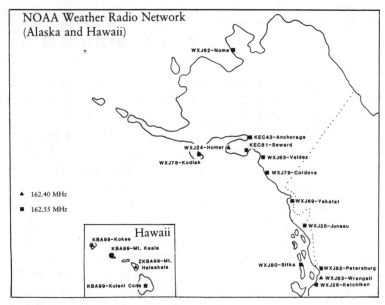

Figure 17.1.

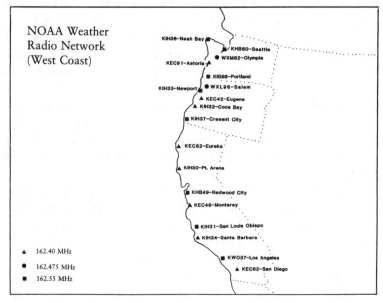

Figure 17.2.

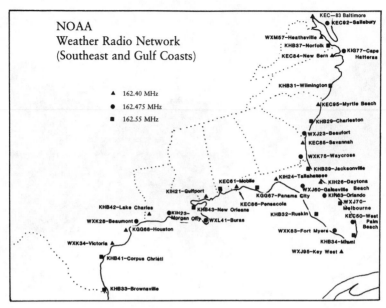

Figure 17.3.

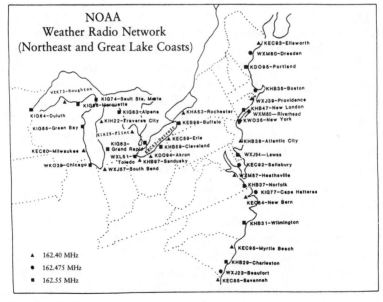

Figure 17.4.

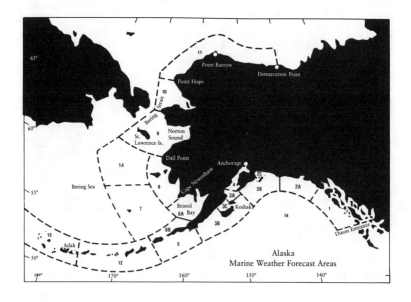

Figure 17.5. 1. *Southeast Alaska, from Dixon Entrance to Cape Fairweather and Skagway; 2A. Cape Fairweather to Cape Suckling; 2B. Cape Suckling to Gore Point; 2C. Prince William Sound; 3A. Kodiak Island waters, Gore Point to Shuyak Island and Cook Inlet south of Chinitna Point and English Bay; 3B. Kodiak Island waters, Shuyak Island to Castle Cape except Shelikof Strait; 3C. Shelikof Strait; 4. Cook Inlet north of Chinitna Point and English Bay; 5. Coastal waters south of the Alaska Peninsula from Castle Cape to Cape Sarichef; 6A. Bristol Bay, north portion from Cape Newenham to Port Heiden; 6B. Bristol Bay, south portion from Port Heiden to Cape Sarichef; 7. Pribilof Islands and Southeastern Bering Offshore; 8. Cape Newenham to Dall Point; 9. Dall-Point to Wales, including Norton Sound and the Saint Lawrence Island waters; 10. Wales to Cape Lisburne, including Kotzebue Sound; 11. Cape Lisburne to Demarcation Point; 12. Aleutian Islands, Cape Sarichef to Adak; 13. Aleutian Islands, Adak to Attu; 14. St. Matthew Island waters.*

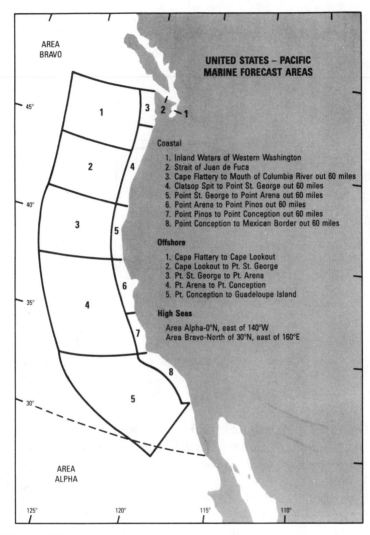

AREA
BRAVO

**UNITED STATES – PACIFIC
MARINE FORECAST AREAS**

45°

1   3   2 ~1

Coastal

1. Inland Waters of Western Washington
2. Strait of Juan de Fuca
3. Cape Flattery to Mouth of Columbia River out 60 miles
4. Clatsop Spit to Point St. George out 60 miles
5. Point St. George to Point Arena out 60 miles
6. Point Arena to Point Pinos out 60 miles
7. Point Pinos to Point Conception out 60 miles
8. Point Conception to Mexican Border out 60 miles

Offshore

1. Cape Flattery to Cape Lookout
2. Cape Lookout to Pt. St. George
3. Pt. St. George to Pt. Arena
4. Pt. Arena to Pt. Conception
5. Pt. Conception to Guadeloupe Island

High Seas

Area Alpha-0°N, east of 140°W
Area Bravo-North of 30°N, east of 160°E

40°

35°

30°

AREA
ALPHA

125°   120°   115°   110°

Figure 17.6.

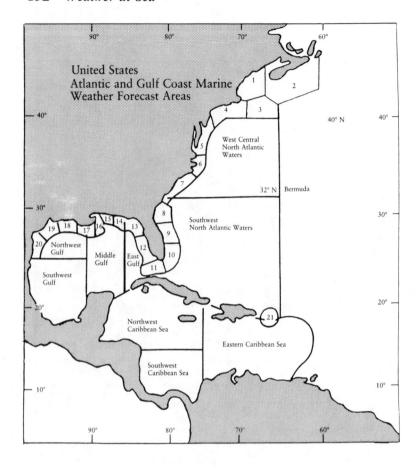

Figure 17.7. *1. Gulf of Maine; 2. South of Nova Scotia; 3. Georges Bank; 4. South of New England; 5. Hudson Canyon to Baltimore Canyon; 6. Baltimore Canyon to Hatteras Canyon; 7. Hatteras Canyon to Blake Ridge; 8. Savannah, GA to St. Augustine, FL; 9. St. Augustine, FL to Jupiter Inlet, FL; 10. Jupiter Inlet, FL to Key Largo, FL; 11. Key Largo, FL to Dry Tortugas, FL; 12. Cape Sable, FL to Tarpon Springs, FL; 13. Tarpon Springs, FL to Apalachicola, FL; 14. Apalachicola, FL to Pensacola, FL; 15. Pensacola, FL to Gulfport, MS; 16. Gulfport, MS to Mississippi River; 17. Mississippi River to Intracoastal City, LA; 18. Intracoastal City, LA to Port Arthur, TX; 19. Port Arthur, TX to Port O'Connor, TX; 20. Port O'Connor, TX to Brownsville, TX; 21. Puerto Rico and Virgin Islands.*

value are the latest hourly (or more frequent) observations from coastal airports and the 12-hour forecast of the wind at 2000 to 3000 feet. These are the levels of low clouds, whose motion is likely to resemble the surface wind offshore, as we pointed out in Chapter 3.

Canada—coastal and adjacent offshore waters. For Canadian waters the Atmospheric Environment Service (AES) provides a continuous weather information service through Weatheradio Canada (WRC), broadcasting at 162.550, 162.400, and 162.475 MHz (corresponding to VHF channels WX1, WX2, and WX3). Warnings, synopses, and forecasts are accompanied by the latest hourly observations from a substantial number of coastal stations. The forecasts, like those of NWR, refer to local coastal waters and a number of appropriate offshore areas, as shown in figures 17.12–17.14. These forecasts describe expected conditions over the following 36 to 48 hours. Continuous Marine broadcasts are issued over VHF

Figure 17.8.

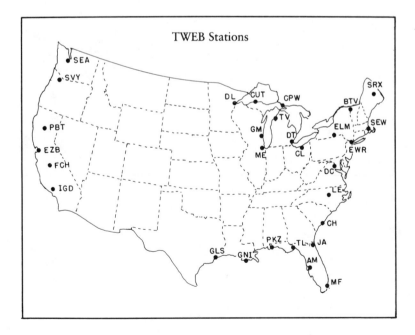

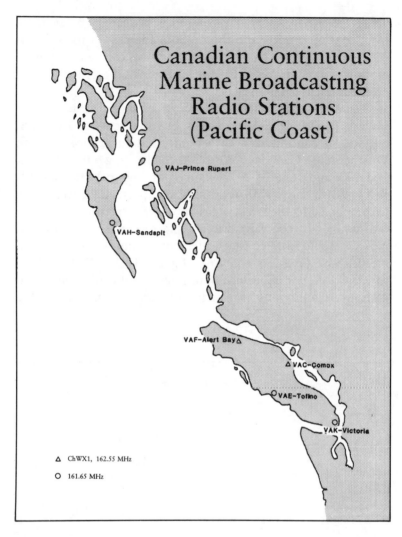

Figure 17.9.

Figure 17.10.

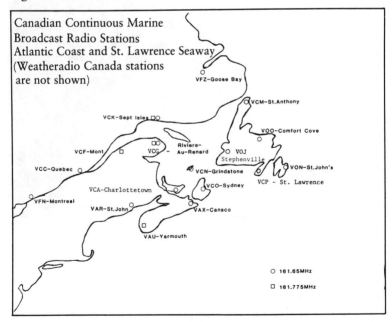

Figure 17.11.

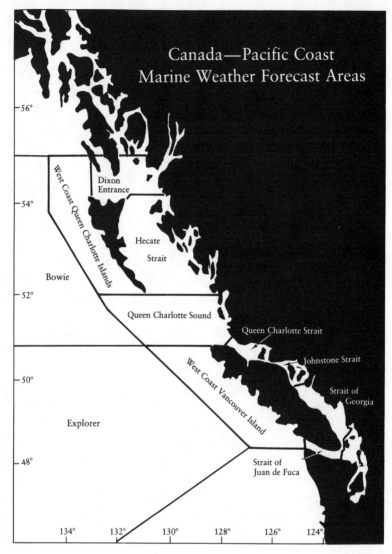

Figure 17.12.

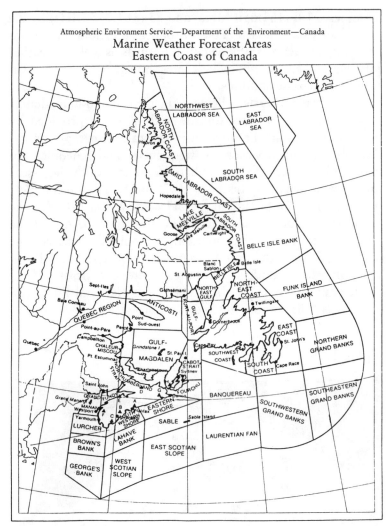

Figure 17.13.

Channels 21 B (161.650 MHz) and 83 (161.775 MHz) by the Canadian Coast Guard. These broadcasts include notices to shipping and reports of navigation hazards as well as weather information, and do not include latest hourly weather observations from coastal stations.

High seas forecasts.   Beyond the coastal and immediately offshore areas, forecasts for the high seas are provided by numerous stations broadcasting via radiotelegraph or radiotelephone on a variety of frequencies and schedules too broad for a specific discussion here. A number of the high-seas areas are shown in figures 17.6, 17.7, and 17.11, and complete information can be found in *Selected Worldwide Marine Weather Broadcasts,* a joint publication of NOAA and the United States Naval Oceanography Command. (This volume also contains information on frequencies, schedules, and contents of radiofacsimile and radioteleprinter broadcasts.)

It is useful to know that WWV, the National Bureau of Standards time station, in Colorado, broadcasts synopses and

Figure 17.14.

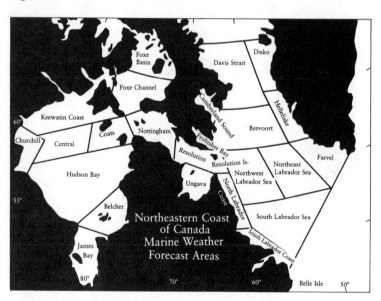

warnings at eight minutes and at nine minutes after each hour, covering the North Atlantic west of 35 degrees west, the Caribbean Sea, and the Gulf of Mexico. This station can almost always be heard on at least one of its single-sideband frequencies: 2.5, 5, 10, and 15 MHz. Similar information for the Pacific coast of North America is broadcast on the same frequencies at 48 minutes past the hour by WWVH in Hawaii.

All of the forecasts we have discussed are renewed at six-hour intervals. Warnings and amendments (if the current forecast has gone seriously awry) are broadcast as and whenever available.

Using the information

There is a great deal more to be gleaned from a weather broadcast than simply the forecast of the wind speed and direction for the particular area in which you are sailing. The information for other areas around, together with the observations from other airports, Coast Guard stations, and buoys are also relevant, and can be used to put together a picture of what the weather is doing farther afield. This picture will enable you to make better judgments concerning trends in the weather, particularly on those occasions when the forecast seems to be going wrong or when you are racing and want to capitalize on every slant of the wind.

In our experience the best way to derive the maximum benefit from weather information is to construct your own weather map for the area, so that you can make the best possible judgment on the meaning and validity of all the data available to you, including your own observations. It will enable you to see at a glance how the major weather systems have evolved, to assess the likelihood of minor troughs and ridges, and to work out what effect the afternoon heating of the land is likely to have on the coastal wind. Unless you know for certain what is the orientation of the geostrophic wind to the coastline, you will not be able to judge the likely character and strength of the sea breeze. For instance, with the geostrophic wind in quadrant four (figures 9.6 and 9.7) will the breeze die away altogether in the afternoon?

A forecaster, when making a forecast for coastal waters, will have in front of him the latest actual chart made at 0000, 0600, 1200, or 1800 UTC (universal coordinated time, another name for Greenwich mean time) and a series of forecast charts covering the period 24 hours or more ahead. You could if you wanted try to use the information broadcast to reconstruct the essential parts of the charts the forecaster started with. It is much easier, however, and just as useful to construct a single weather map for the time for which you have observations. This will mean interpolating the appropriate positions of lows and highs from those given in the discussion and forecast. The steps to this reconstruction are as follows:

1. Transcribe as much of the weather broadcast as you can manage. Some people find it a great help to record the broadcast on tape.
2. Take a blank map of the area. You will have to prepare a supply of these in advance.
3. Write down the values of pressure, wind, and visibility at the positions of each of the observing points, and make sure you also write somewhere on the chart the date and time the observations were made. Do not overlook including your own observation as well; it is just as important as the rest.
4. Plot the positions of the highs and lows, and their intensities or pressures if given, from the discussion. Also sketch in the positions of fronts and other troughs and ridges that may be mentioned. Note that your intent is to construct a weather map for the time of your observations. This means interpolating the positions of the various map features between the starting time of the discussion and the ending time of the forecast.
5. Write in each coastal or offshore area the wind speed and direction taken from the forecast, but appropriate to the time of the observations.

6. You now have a lot of wind information on your chart and hopefully a few pressure values. From these, sketch in the isobars to fit the pressure values, aligned so that the wind is blowing along them but slightly inward toward lower pressure and outward from high pressure, and so that their distance apart is inversely proportional to the wind speed. It is best to use standard values for the isobars starting from 1016 millibars (30.00 inches) and at intervals of 4 mb (0.12 inch), or 2 mb if the wind is light. For a lead-in to your first map, a sequence of maps cut out from a newspaper or your own rough sketches from television broadcasts will be a great help. Yours will be the next one in the series.

7. Do not forget that winds reported by coastal stations may be influenced by local coastal effects and land and sea breezes, so that they may not fit the isobars you draw. You will have to work it out. Winds from buoys and small islands far from the mainland are more reliable.

Terms used in weather broadcasts

Wind.   The direction is given according to the eight-point compass, and the speed in knots. The description "gale" is used for winds between 34 and 47 knots and "storm" for 48 knots and above.

Sea state.   The sea state is described in terms of the significant wave height in feet. This is the average of the highest one-third of the waves experienced. Remember, however, that this is an average value and that a few individual waves substantially higher than this must be expected. Information on swell may also be included, in which case its direction of propagation is given.

Weather.   Only "significant" weather is included: rain, thunder, fog, etc. The amount of cloud is not normally mentioned.

Visibility.   This is expressed in miles in the United States but may not be specifically mentioned unless it is three miles or less. It is expressed in meters or kilometers in Canada. The term *haze* may be added in the case of visibilities between two and five miles, and *fog* for lower values.

Icing and sea ice.   Information on icing refers to the possibility of freezing precipitation or freezing spray, both of which are dangerous to the stability of the boat.

Transcribing weather information

Even if you transcribe the data in longhand you will still need to use some form of abbreviation to put the information on your weather map. There is much to be said for using the same shorthand throughout. There is nothing to stop you from devising your own shorthand, but it is a good idea to use a notation that has been developed by people with considerable experience in using weather broadcasts and which includes a number of the standard international weather map symbols. Once you are familiar with the more common of these you will be able to appreciate at a glance the information contained on any weather map you may see displayed in clubs or at ports of call. The common international plotting symbols are shown in the accompanying chart.

The synopsis

Few international symbols are involved in the reporting of the current positions and movements of systems except for the points of the compass—N, S, NW, SW, etc. Capital letters are the best shorthand for weather systems—L for low, H for high, T for trough, and R for ridge. One very useful hint is to use a vertical stroke (or slash) to denote the passage of time.

Weather	letter	plotting symbol
rain	r	●
drizzle	d	,
snow	s	✳
shower	p	▽
hail	h	△
thunderstorm	th	ꝉ
squall	q	∀
mist	m	=
fog	f	≡
haze	z	∞

The headline

The headline comprises a summary of all warnings and advisories in force, drawing attention to any hazardous conditions contained in the forecast.

Sea area forecasts

Wind direction is always given in terms of the eight-point compass and the speed in knots. Using the slash for the passage of time, a sentence such as, "Northwest 20 to 25 knots at first, backing to southwest and increasing to 35 to 40 knots at the end of the period," is written simply as "NW 20–25 / SW 35–40." Similarly, "from the south at first" is written "S /" and "from the north later" as "/ N."

Weather

Weather is always given in terms such as "clear," "cloudy" (perhaps with a qualifier), "rain," etc. International shorthand should be used, and you can choose between the letter notation and the international weather map symbols as given above.

There is something to be said for using the symbols because you can then plot these directly on your map, but the former are much easier to learn, and you can readily turn them into plotting symbols at your leisure after the broadcast. When a change in the character of the precipitation is forecast to occur during the specified period, such as "rain, tapering off to showers," you can write it "r/p." This sort of detail should always be taken down, since it almost certainly ties in with a change of wind and the passage of an important weather system through the area.

For heavy precipitation, capital letters are used; for example, R connotes heavy rain.

Station reports

As with the sea area forecasts, when transcribing station reports you need a prepared form with the station names listed, whether they are airports, Coast Guard stations, or buoys. The form should have columns for the following:

• wind
• significant weather
• visibility
• barometric pressure
• wave height
• remarks, such as "breaks in the overcast," "fogbank to the SE," "distant lightning to the NW," "rain occasionally heavy," "pressure falling rapidly," and so forth.

The same shorthand should be used as for the forecasts.

——————— Drawing a weather map

Having taken down the information, your next step is to plot it on a weather chart. Throughout the world there is a standard format for doing this. The wind is always drawn as a feathered arrow blowing toward the observing position, and with the "feathers" always on the left as you stand with back to wind. That is, the feathers point in the general direction of lower pressure. The number of feathers is always proportional to wind speed, one feather for each ten knots and a half-feather for five knots. The rest of the information is placed around the observing position like this:

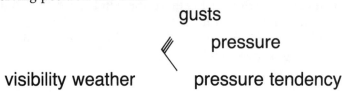

gusts

pressure

visibility weather   pressure tendency

So "northwest 25 knots with gusts to 35 knots, continuous rain, 1009 mb, pressure falling, visibility two miles" is plotted like this:

G35

09

2

Similarly, plot the wind and the weather in each one of the coastal and offshore areas, as appropriate for the amount of information given in the forecast. For example, a forecast of "northwest 25 knots in the east and 35 knots in the west, with visibility occasionally 3 miles in showers" is plotted:

3   3

When ranges are given it is best to plot the worst value—the highest in the case of wind (but remember that you have done so when fitting the isobars), and the lowest in the case of visibility.

Drawing the isobars

We know that the pressure gradient, which on a weather map is given by the distance between isobars, is inversely proportional to wind speed, other things being the same. So relating wind speed to the distance between isobars we can either derive the wind speed from isobars already drawn or draw the isobars at the right spacing given the wind speed. We must of course take into account the Coriolis force, which varies with latitude. A useful rule is that for a 20-knot geostrophic wind, the spacing of two-millibar isobars is as follows:

Latitude	Spacing (nautical miles)
60	62
55	66
50	70
45	76
40	84
35	94
30	107
25	128
20	157

Referring if possible to a previous weather map, either in the newspaper or one you drew yourself, draw the new positions of the isobars starting where you have observed pressure values and then extending outward to the known positions of centers of low and high pressure. Over sea areas, the isobars should be drawn almost parallel to the wind, so that the wind is blowing slightly toward lower pressure. Keep to standard values for the isobars, usually every 4 mb up or down from 1016 mb, or in light winds every 2 mb.

Drawing the centers, troughs, and fronts

Here considerations of continuity are particularly useful. Where was the front 24 or 12 hours ago? Where is it forecast to be? How far is it likely to have moved from the last known position to the time for which you are constructing the chart, given the forecast?

Finally, having estimated the position of the trough or front, identify it with the standard symbols: a dashed line for a trough, and for a front a solid line with blobs or spikes protruding from its leading edge depending on whether it is a warm or cold front. You may then want to redraw your isobars to indicate the degree and sharpness of the veering wind shift at the front or trough, if you know something about it.

_____ Using your weather map

Having drawn your weather map, you will have a much better understanding of the weather situation and a much better ability to interpret the naked forecast than if you had just written it down for your sea area. For instance, the forecast wind may be southwesterly at 20 knots. It may be critical for a long beat to know whether "southwesterly" means 220 degrees or 240 degrees. Your map will help you decide. It will also help you decide whether 25 knots is more likely over one side of the forecast area than the other, or later rather than sooner. When the direction is given as "southwest to west" your map may reveal a bend in the isobars but no particular moving feature, or it may show a weak trough which you can expect to be moving through the area. A multitude of minor influences are possible, which, taken over a season, make your sailing safer, more skillful, and more enjoyable.

Figure 17.15. Following three pages: *These weather maps were plotted and drawn by the recommended methods, from broadcast information received by one of the authors from Weatheradio Canada while cruising in the Canadian Maritimes. The observations made by the author were especially useful to him in these cases, representing the only local pressure readings available. The plotting convention is the one illustrated in this chapter, except that three digits are plotted for observed pressure—that is, 198 means 1019.8 millibars. Also, a solid dot above an inverted triangle signifies rain showers; two side-by-side solid dots denote continuous light rain; a triangle of three solid dots means continuous moderate rain; and four solid dots in a diamond-shaped array indicate continuous heavy rain. A*

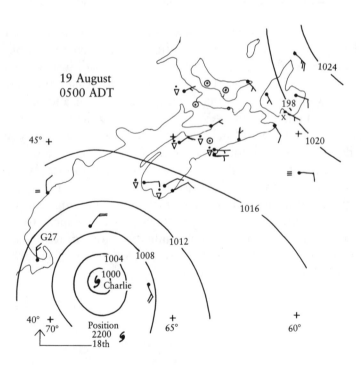

capital "G" followed by a number gives the local strength of the gusts. A solid dot enclosed in a circle indicates a non-reporting station. The crosses are lat/long intercepts for 5-degree increments of latitude and longitude.

The first two maps illustrate the eastward passage of a tropical storm well to the south of Nova Scotia that nevertheless produced gales reaching the coast. The yacht (marked with an X) was anchored in a cove on Cape Breton Island, open to the northwest, in the correct anticipation that strong winds would not blow from that direction. The other two maps illustrate the passage of an active cold front across Nova Scotia, preceded by southerly gusts of gale force. The yacht in this case was anchored just north of a small island along the south coast of Nova Scotia, for shelter from the predicted southerly gale.

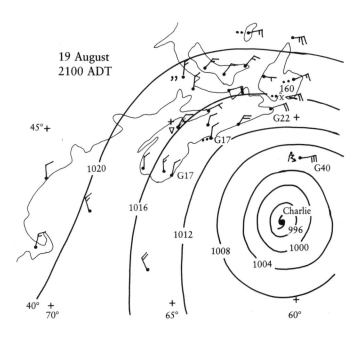

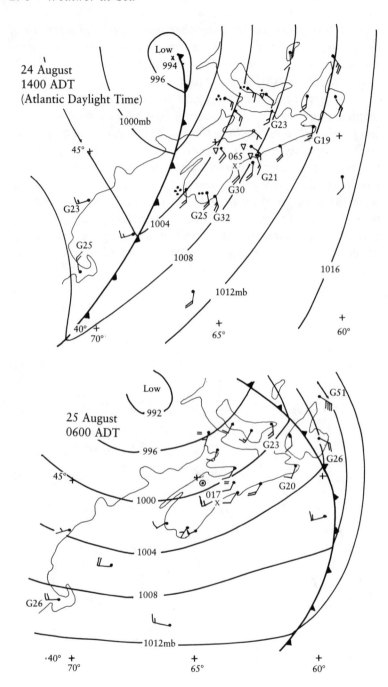

Using the observations

It is always useful to make a special study of the observations available from places close to where you are sailing.

If you are in a rapidly changing weather situation, it is the observations upwind that are most important, and you need to look carefully at those up to 200 miles away, if not more, for signs of what may happen in the next few hours.

If the weather is slow-moving, particularly if it is anticyclonic and the wind is ten knots or less, careful attention to all observations within a hundred miles or so will pay dividends. Plot them all on your weather map and follow them through in time sequence, noting the changes in pressure, wind direction, wind speed, and visibility. Make allowance for the "tidal" effect in pressure values and for night and day coastal effects on the wind, such as land and sea breezes. You are very likely to identify small changes or bends in pressure gradient and in wind too slight to be mentioned in the forecast. For instance, observations from offshore points may show eddies in the wind on the order of 5 to 15 miles across. Never say, "I don't believe it—it doesn't fit." You may be right, but on the other hand it may signify something important that neither you nor the forecaster knew about before this. So think twice!

Weather Questions

Q. Why are there more serious thunderstorms on the East Coast than on the West Coast?

A. *It is a matter of the water temperature, which is much higher, latitude for latitude, on the East Coast than on the West Coast. See Chapter 15, page 133.*

Q. Why do gales tend to be more severe in the colder season?

A. *The intensity of mid-latitude cyclones depends on the strength of horizontal temperature contrasts between low and high latitudes, which is much greater in the cold season. See Chapter 5.*

Q. Why on the East Coast does a northwest breeze usually die away at sunset in summer, and spring up again the next morning?

A. *The strong cooling of air next to the land surface at night prevents the continuing breeze aloft from reaching the ground. See Chapter 8, page 89.*

Q. Why are there sometimes open patches in the lee of islands in a fogbank?

A. *Over land, the turbulent mixing due to wind and the day-time heating tend to dissipate the fog, a benefit extending to water in the immediate lee of an island. See Chapter 15, page 143.*

Q. Why is local wind frequently so different from that reported in the weather broadcasts?

A. *The features of shorelines with bays and islands, and the
character of the inland area, produce variations of surface
friction, and of heating and cooling, that are responsible
for variability in the wind as great as that in the terrain
itself. See Chapters 8 and 9.*

Q. When are large-scale weather reports and forecasts most
likely to be applicable with little or no modification from
one locale to the next?

A. *When the large-scale wind systems are strongly developed,
there is relatively little opportunity for local circulations to
develop, and even those that are embedded attract less
notice. See Chapters 8, 9, 10, and 13.*

Q. What is the cause of the famous "nor'easterly" gale?

A. *This is the wind found north of a northeastward-moving
cyclone along the East Coast, often enhanced by the
damming of the northeasterly flow of cold air by the mountains to the west. See Chapters 5, 7, and 8.*

Q. Can a turn of the tide have any effect on the breeze?

A. *The accompanying change of water temperature may on
occasion bring about an increased mixing of the breeze
aloft down to the surface, relieving a frustrating calm. See
Chapter 11, page 109.*

Q. How can you tell when fog is imminent? How can you tell
when it is likely to scale up?

A. *When the dew-point temperature of the air exceeds the
temperature of the water, fog is likely sooner or later. Near
shore, daily heating over peninsulas and offshore islands
will break up the fog there and often over waters to the
immediate lee. Listen to broadcasts of the latest coastal*

observations and keep a sharp watch. See Chapters 2, 15, and 17.

Q. Is there any way to tell whether a local thunderstorm will produce squalls?

A. *When heavy rain is falling, the accompanying downrush of cooled air will spread out well in advance of the rain shaft, sometimes accompanied by a roll cloud. See Chapters 6 and 15.*

Q. Is there any way to know whether a 25-knot breeze is likely to get stronger?

A. *If it's a local sea breeze, it probably won't get any stronger, especially in late afternoon. If all the signs indicate the approach of a cyclone, stay on top of the forecasts and be aware whether a gale warning is in effect. See Chapters 4, 9, 15 and 17.*

Q. How useful are clouds as weather predictors? Can they offer details about local weather not available from the large-scale broadcasts?

A. *They are useful in many ways. The first sign that a sea breeze will develop is the appearance of an array of small cumulus clouds just inland along the coast. Thunderstorms may be mentioned in the forecast, but with no specifics as to time and place. Your observations are most helpful here. See Chapters 6, 9, and 15.*

Q. With their computers and satellites, why aren't the weathermen always right?

A. *If weather systems were so simple that we could be right all the time, we wouldn't need computers. Tide predictions were highly accurate long before the computer age. Satellite photographs show how complex atmospheric conditions are at any one moment. Two computer forecasts started from the same moment will diverge if the starting* *conditions are different to the slightest degree, far beyond our ability to distinguish. See Chapters 1 and 5.*

Q. Is there any logical repeating pattern for the way a sea breeze fills in over a particular bay that would be helpful to me in local races?

A. *The variety of details in the development of a sea breeze matches the variety in the local coastline and topography, but the general principles are always the same. See Chapter 9.*

Q. Why does the dominant East Coast pattern of southerly breezes in the summer give way to faster-moving systems and more variable winds the rest of the year?

A. *Because of the stronger north-south temperature contrasts during the cold season, the cyclones and anticyclones are more vigorous and strong fronts are more likely in the colder seasons. In summer the sea breeze dominates during the daytime and the passing weather systems are weaker and more sluggish. See Chapters 5 and 9.*

Q. Why does a summer sea breeze typically not fill in until the afternoon, then die away toward sunset?

A. *The sea breeze is closely tied to the heating of the land by the sun while the sea surface temperature changes hardly at all. The resulting circulation follows the heating cycle, but with some time lag. See Chapter 9.*

Q. Why is there often more wind at the masthead than at deck level?

A. *Friction brings the wind to a literal halt exactly at the sea surface. When mixing of air along the vertical is inhibited, the contrast between deck and masthead is strong. See Chapter 3, page 33.*

Q. Does wind always accompany rain?

A. *As a rule, the rising of air that leads to rain is part of a circulation system that necessarily includes horizontal air motion, or wind, whether the system is a cyclone or a thunderstorm. Local calm patches can be part of the system, however. See Chapters 2, 4, 5, and 15.*

Q. In a light breeze, is there less wind in the patches where no wind ripples appear?

A. *As a rule, smooth patches on the water surface correspond to areas of especially weak wind between areas of somewhat stronger wind, even over the open ocean. Sometimes, however, when the water is much colder than the air, there can be plenty of wind at the masthead above perfectly smooth water. See Chapters 3 and 10.*

Q. Why does the weather sometimes clear up dramatically with a whipping northwester, while at times it only "ceases raining," and the clearing breeze is delayed?

A. *A sharp cold front with strong descending motion in the cold air will bring strong wind and sudden clearing, while the rain well ahead of a low tracking eastward to the north of you will give way to a period of mild air with little vertical motion before the cold front arrives. See Chapters 4, 5, 6, and 7.*

Q. Why do northwesters seem to be the strongest at the start, then taper off?

A. *When the wind first shifts to northwest, the cyclone center is nearby and the isobars are closely packed. Then, as the next anticyclone approaches, the isobars become more widely spaced and the wind slowly dies. See Chapters 3 and 4.*

Q. Does rain calm the seas?

A. *The impact of raindrops probably disrupts the patterns of the smaller high-frequency waves in the spectrum, which* *may have a subsequent effect on the development of the larger, low-frequency waves. But the effect is typically masked by the association of rain with generally windy conditions.*

Q. Why is fog more likely to be accompanied by light breezes than heavy ones?

A. *Fog forms when the air is cooled from below. This cooling also inhibits mixing of strong wind aloft down to the surface. See Chapters 2, 3, and 15.*

Q. Why are west and northwest winds on the East Coast often gustier than southwesterlies and southeasterlies?

A. *A northwesterly is a land breeze, filled with turbulent gusts that reach the surface because of the strong heating of the air over land. Southerlies are typically being cooled from below, especially over the cold water of the Northeast, so that the stronger winds aloft are inhibited from reaching the surface. See Chapters 2 and 3.*

Q. What causes the strong northerlies that last for several days off the coasts of Oregon and Northern California during the summer?

A. *Intense heating over the deserts of the Southwest, contrasted with the coldness of coastal waters, leads to close spacing of isobars from a low-pressure trough on the coast to the elevated pressures of the Eastern Pacific High. This situation can produce 30-knot winds and 14-foot seas that last for days. Similar heating contrasts on a much smaller scale produce the daily sea breeze. See Chapters 2, 3, and 9.*

Q. Why do most sailors headed for Hawaii sail considerably south of the rhumbline when departing ports in the Pacific Northwest?

A. *A rhumbline course would take you through the light winds in the center of the Eastern Pacific High, a highly*

*persistent feature of the summer pattern. The greater dis-
tance on a course farther south is more than compensated
by the better wind. See Chapters 1, 3, and 4.*

Q. Why do we have more fog on the West Coast in the summer than during the winter?

A. *Fog at sea occurs where the dew-point temperature of the
air is higher than the water temperature. Since the sea-surface temperature changes relatively little from summer to
winter, while the air is much drier in the winter, the poten-
tial for fog here, and on the East Coast as well, is greater
in summer. Along the Gulf Coast fog is more frequent
in winter, when the air remains warm but the water is
cooled by the discharge of seasonally swollen rivers. See
Chapter 15.*

Q. What causes typical summer winds in Southern California
to change from light southwesterlies in the morning to
westerly with peak speeds in the late afternoon, followed
by a return to southerly?

A. *This behavior is caused by the intense heating over land,
with relatively cool water offshore. The process resembles
a classic sea breeze, but doesn't include the return flow
aloft. See Chapter 9, page 99.*

Q. Why do waves in the Gulf Stream seem especially dangerous?

A. *It is generally rougher in the Gulf Stream than in the cooler
water on the Continental Slope, because the warmth of the
water brings strong wind immediately close to the surface.
Gulf Stream waves are particularly dangerous when the
current is running against the wind. See Chapters 3 and 12.*

Q. Why is there much more sea swell on the West Coast than
on the East Coast?

A. *The swell on the West Coast represents waves made by
storms over a vast expanse of ocean to the west. Storms in*

*the western Atlantic send swell eastward and southward
from strong generating regions of westerly and northerly
winds, but only rarely westward. See Chapter 12.*

Q. Why do hurricanes die so quickly when they cross the
coastline?

A. *Tropical storms are crucially dependent on the latent heat
released when massive amounts of water vapor condense
into cloud and rain. This supply can be provided only by an
extremely warm sea surface. See Chapter 14.*

Q. Why are winds stronger on the right-hand side of the path
of a hurricane?

A. *Hurricanes move with the large-scale wind flow in which
they are embedded. The circular wind vortex due to the
storm is added to this large-scale wind on the right side and
subtracted on the left. Mid-latitude cyclones do not follow
this rule, because their motion results from different
causes. See Chapters 5 and 14.*

Glossary

Adiabatic. Characteristic of a process in which there is no exchange of heat between an imaginary parcel of air and its environment. The changes of temperature that occur in an adiabatic process are due solely to the expansion or compression of the air as its pressure changes, usually because the air is moving up or down. Transfer of heat by radiation or by mixing with adjacent air are examples of nonadiabatic processes.

Adiabatic Lapse Rate. The rate at which a parcel of air will cool while its pressure is falling (or warm while its pressure is rising) when no heat is being exchanged between the parcel and its environment. In the usual case, when the pressure change is due to ascent or descent of the parcel, the adiabatic lapse rate is very nearly 5.5°F for every thousand feet of change in elevation.

Advection. A process in which the wind flow tends to produce a change in some weather element, usually temperature or humidity, at a fixed point. If at some point, for example, the wind is blowing across the isotherms from colder toward warmer air, the process of cold advection is occurring, and if no other process is acting with sufficient strength to balance it, the temperature will fall at that point. The strength of the advection is the product of the speed of the wind and, in this example, the strength of the variation of temperature along the direction of the wind. The strength is inversely proportional to the spacing of isotherms on the map, measured along the direction of the wind.

Anemometer. An instrument for measuring wind speed and direction. In a typical masthead installation, three or four rotating cups measure speed, while a vane indicates direction.

Anticyclone. A system of winds rotating around a central point, in a clockwise sense in the northern hemisphere. The barometric pressure is a maximum at the center.

Back. A change of wind direction in the counterclockwise sense. The term is used to describe the change with time at a given point or the change either with elevation or in a horizontal direction at a given time.

Barometric Pressure. The pressure exerted on a surface by the atmosphere. It is very close to the weight of the air in the column overhead—thus the higher you go, the smaller the barometric pressure. In order that small but important horizontal changes in pressure at a fixed level can be estimated from a set of observations at different elevations, the observed values are typically corrected to refer to some fixed elevation, usually mean sea level. The pressure is measured by one of a number of kinds of barometer, the two most common being the mercury-in-glass type and the aneroid type. The first measures directly the pressure on a free surface of mercury by making it equal to the weight of a column of mercury in a closed tube from which the air has been removed. The height of this column at sea level is typically about 30.00 inches. This is the equivalent of a pressure of 1016 millibars, the physical units used most often by meteorologists. The aneroid type of barometer measures the deflection of the end walls of a cylindrical shell from which the air has been removed. This deflection depends on the barometric pressure outside, so the pressure itself is not directly measured and the instrument must be calibrated. Aneroid barometers are much more compact than mercury ones and are much more widely used.

Beaufort Wind Scale. An arbitrary numerical scale of wind force devised by Admiral Francis Beaufort in the mid-19th century. It was framed originally in terms of the amount of sail that could be carried safely by a full-rigged ship. Ranges of wind speed corresponding to the Beaufort numbers and appropriate descriptions of the sea state and of indicators on land

were added later. It is still widely used in British marine forecasts and observations. A modern version of the Beaufort scale
is shown here.

Beaufort scale of wind force				
Beaufort No.	General Description	At sea	On land	Limits of velocity in knots
0	Calm	Sea like a mirror.	Calm; smoke rises vertically.	Less than 1
1	Light air	Ripples.	Direction of wind shown by smoke drift but not by wind vanes.	1 to 3
2	Light breeze	Small wavelets.	Wind felt on face; leaves rustle.	4 to 6
3	Gentle breze	Large wavelets. Crests begin to break.	Leaves and small twigs in constant motion. Wind extends light flags.	7 to 10
4	Moderate	Small waves becoming longer, fairly frequent white horses.	Raises dust and loose paper; small branches are moved.	11 to 16
5	Fresh breeze	Moderate waves, many white horses, chance of some spray.	Small trees in leaf begin to sway.	17 to 21
6	Strong breeze	Large waves begin to form; the white foam crests are more extensive everywhere. Probably some spray.	Large branches in motion Umbrellas used with difficulty.	22 to 27
7	Near gale	Sea heaps up and white foam from breaking waves begins to be blown in streaks along the direction of the wind.	Whole trees in motion	28 to 33
8	Gale	Moderately high waves of greater length; edges of crests begin to break into spindrift. The foam is blown in well-marked streaks along the direction of the wind.	Breaks twigs off trees; generally impedes progress	34 to 40
9	Severe gale	High waves. Crests of waves begin to topple, tumble and roll over. Spray may affect visibility.	Slight structural damage (chimney-pots and slates removed).	41 to 47
10	Storm	Very high waves with long overhanging crests.	Seldom experienced inland trees uprooted, considerable structural damage occurs.	48 to 55
11	Violent storm			56 to 63
12	Hurricane			Greater than 63

Buoyancy. The difference in density between a parcel of air,
such as the air in a cumulus cloud, and the air in its environment. If the parcel is less dense than the environment, positive
buoyancy will cause it to accelerate upward. If it is more dense,
negative buoyancy will cause it to accelerate downward.

Col. In a pattern of pressure, a point where a ridge of high
pressure and a trough of low pressure intersect. It is analogous
to a saddle point, or mountain pass, in the topography of land
elevation—the highest point in a valley and also the lowest
point in a ridge.

Cold Front. A front, typically extending southwestward from a
cyclone center, which is moving across the map so that the
colder air is advancing toward the region of warmer air. Thus

if you are at rest at a particular point on the sea surface, you will experience a sudden drop of temperature accompanied by a clockwise wind shift, typically from southwesterly to northwesterly, as the cold front passes. One or more bands or squalls of rain are often experienced before, during, or after passage of the front.

Condensation. The process of change from water vapor to liquid water—that is, cloud, fog, or dew.

Coriolis Force. In meteorology, an apparent force acting on a parcel of air, due to the rotation of the earth about its axis. It is a maximum at the poles and drops gradually to zero at the equator, acting to the right of the direction the air is moving in the northern hemisphere and to the left in the southern hemisphere. At a given latitude, the strength of the Coriolis force is proportional to the wind speed. Its greatest practical effect is that the wind as we measure it, above the lowest levels where surface friction is important, blows parallel to the isobars.

Cumulus Clouds. Clouds that form when a small mass of air rises and cools, with condensation of water vapor into cloud droplets, through an environment that is cooler than the rising air parcel. Buoyancy then encourages further growth of the cloud, which may develop further into a shower or thunderstorm.

Cyclogenesis. The process of formation of a cyclone. A center of lower pressure forms at the same time.

Cyclone. A system of winds rotating completely around a central point, in a counterclockwise sense on the northern hemisphere. A minimum of barometric pressure occurs at the center.

Dew Point. The lowest point to which a sample of air can be cooled, without changing its water-vapor content or pressure, before condensation begins. When the temperature of air is equal to its dew point, the air is saturated with water vapor.

Diurnal Cycle. The variation of temperature, wind, or any other weather element over the course of a day.

Environmental Lapse Rate. The rate of fall of temperature with elevation in a given air column at a particular time. It is measured instantaneously but roughly by the infrared sensing system of a satellite, or to a close degree of approximation but not quite instantaneously by an instrumented balloon, rising at about 10 mph.

Evaporation. The process of change from liquid water to water vapor.

Front. A narrow elongated zone of pronounced temperature contrast accompanied by a wind contrast in the cyclonic sense —for example, southwesterly in the warmer air and northeasterly in the colder air. One or more bands of cloud and rain typically accompany the front. A front forms when the wind pattern on the broad scale is so configured as to transport warmer and colder air separated by about a hundred miles into close proximity in about a day. A trough of relatively low barometric pressure accompanies the front, and a low center, called a frontal cyclone, often is found in the trough. The low-pressure trough develops simultaneously with the front. Either the front or the low center may appear first. *See also* Cold front; Occluded front; Warm front.

Frontal Discontinuity. The extremely sharp contrast of wind and temperature over a distance of a mile or less that can occur when a frontal system is strongly developed.

Frontal System. A term referring to the ensemble of typical characteristics that accompany a front, usually emphasizing the clouds and rainfall.

Frontal Trough. A sharp trough of low pressure that accompanies a front. The horizontal pressure gradient changes abruptly across the frontal trough, and so does the wind.

Frontogenesis. The process in which the contrast between warmer and colder air is sharpened nearly to a discontinuity. This development is typically accompanied by a sharp wind shift and by a band of rain falling from the warmer air along the frontal line.

Gap Wind. A strong local wind blowing through a narrow gap in a mountain range, such as occurs along much of the West Coast of North America. It is important for sailors when it blows from the interior out over the sea.

Geostrophic Wind. Sometimes called the "gradient wind." An ideal wind flow in which the Coriolis force acting on the air parcel is exactly balanced by the horizontal pressure-gradient force. Since the Coriolis force acts at right angles to the wind direction, the pressure-gradient force must also be at right angles to the wind direction for a balance to be possible. The wind then blows along the isobars. In the northern hemisphere the Coriolis force acts to the right of the wind direction, so you will find lower pressure to the left if you stand with your back to the wind. Above the shallow layer where surface friction is important, the actual wind is usually close to the geostrophic ideal, and at sea the wind on your sail is usually closely related to the geostrophic direction and speed.

Gradient Wind. A hypothetical curving wind flow in which there is a balance between the pressure-gradient force, the Coriolis force, and the centrifugal force. It is sometimes erroneously used as a synonym for geostrophic wind, which represents a balance between just the pressure-gradient force and the Coriolis force.

High. In meteorology, a region where the barometric pressure is high relative to its horizontal surroundings. *See* Anticyclone.

Inversion. An environmental lapse rate in which the temperature increases with increasing elevation. This is the inverse of the usual situation, but is not uncommon in the lowest layer of the atmosphere when the air is being cooled by the surface, either over cold water or over land on a calm, clear night.

Isobar. A line on a surface, such as a weather map of conditions at a fixed elevation, representing a selected value of barometric pressure. The line separates regions on this surface where the pressure is higher than the selected value from regions where it is lower. Along the line, the pressure is exactly equal to the selected value. To describe the pressure pattern on a surface, isobars are shown for a number of selected values, typically differing by a constant interval, as for example 1016 millibars (30.00 inches) and every two millibars (.06 inch) above or below this value. When the gradient of pressure along a surface is strong, the isobars are closely spaced.

Isotherm. The counterpart of isobar, except that it refers to the temperature. We often speak of the isotherms of mean temperature in a layer between two specified elevations.

Jet Stream. A strong, narrow, elongated current of wind in which speeds during winter can exceed 200 knots. Its core is usually found at elevations between 30,000 and 40,000 feet, but its lower portions can often be traced down to elevations near 15,000 feet. Much weaker, separate, low-level jets sometimes occur below 5,000 feet.

Land Breeze. A gentle breeze that blows from land to sea near shorelines at night when the land is cooled strongly while temperatures over the sea remain nearly constant. It is strengthened and channeled by mountains, hills, and even minor undulations of the terrain.

Lapse Rate. The rate of fall of temperature with elevation. This may refer to the change experienced by a single identifiable air parcel as it rises (adiabatic lapse rate) or to the vertical profile of ambient temperature at a given instant (environmental lapse rate).

Latent Heat. The heat released by water vapor when it condenses into liquid form, or the heat required by the liquid to evaporate it. In the atmosphere, the condensation is in the form of cloud droplets that may subsequently combine with other droplets or with ice crystals to form snowflakes or raindrops. These particles often evaporate before reaching the ground as precipitation, when they fall from the parent cloud into drier air below. The release of latent heat can be an important warming effect in the atmosphere, and the cooling by evaporation of rain is important for the generation of squalls in thunderstorms.

Low. In meteorology, a region where the barometric pressure is low relative to its horizontal surroundings. *See* Cyclone.

Millibar. One one-thousandth of a bar. A millibar is the pressure exerted by a force large enough to accelerate 100 kilograms of mass at rest to a speed of one meter per second, in one second, acting on an area of one square meter. In more applicable terms, the weight of an average column of air overhead at sea level exerts a pressure of 1016 millibars, which is the equivalent of the weight of a column of mercury 30 inches high.

Moist Adiabatic Lapse Rate. The temperature change experienced by a parcel of air due to pressure change, usually caused by ascent or descent, when the process is adiabatic except for the addition or subtraction of latent heat when the air is saturated and either water vapor condenses or suspended water droplets evaporate. The moist adiabatic lapse rate is always smaller in magnitude than the ordinary adiabatic lapse rate, but approaches it when the air is very cold so that the water-vapor content is extremely small.

Nimbostratus. A flat, featureless layer of cloud from which rain is falling.

Nimbus. In meteorology, a cloud from which rain is falling. If the cloud is a cumulus that has developed to great heights, with heavy rain and violent updrafts and downdrafts, it is called a cumulonimbus.

Occluded Front. A narrow zone of wind shift, typically from southeasterly to southwesterly, often associated with a band of cloud and rain of variable width, which is the result of the disappearance of a previous warm sector due to the overtaking of a warm front by a faster-moving cold front as they both rotate around the center of a cyclone. Similar structures, with a wind shift and band of clouds and rain but no notable significant temperature change, can form without a previous warm sector and frontal structure.

Offshore Breeze. A gentle wind near a coast, blowing from land toward the sea. A land breeze is one instance of an offshore breeze.

Onshore Breeze. A gentle wind near a coast, blowing from the sea toward land.

Orographic. Referring to mountains or hills and the valleys between them. Orographic effects on wind and rainfall can be very pronounced.

Pressure. *See* Barometric pressure.

Ridge. Referring to pressure, the extension of a region of relatively high pressure away from a center, analogous to a mountain ridge.

Saturation. In meteorology, the state of a sample of air in which the water-vapor content is as large as it can be without condensation occurring. Saturated air at sea level with a temperature of 70°F will have a water-vapor content of 1.6 percent by weight. This percentage changes by a factor of two for each temperature change in saturated air of about 19°F, becoming smaller with cooler air.

Sea Breeze. A wind blowing during the daytime from the sea, or from a large lake, across the coast line. It is the result of strong heating of the land surface, but only the slightest heating of the sea surface, by the sun.

Squall. A strong wind that arises suddenly and lasts only a short time. Most squalls are associated with thunderstorms or cold fronts.

Stable. A condition in which an individual parcel of air if displaced upward becomes colder than its environment, or if displaced downward becomes warmer than its environment. Because of the density differences between the parcel and its environment, it tends to return to its original height in either case.

Stationary Front. A front that is moving at an insignificantly small speed. The greatest frontal intensity often develops when the front is stationary.

Stratocumulus. A layer of cloud that displays a pattern of darker and lighter spots, showing that it is thin in some places and thick in others.

Stratus Clouds. Layered clouds that form when the entire layer of air is cooled, usually adiabatically during ascent associated with a cyclone, a front, or flow over a range of mountains. The term is often restricted to such clouds below about 6000 feet elevation, although clouds of stratus form can occur at any level up to 45,000 feet or so.

Thermal Wind. The difference in the direction and speed of the geostrophic wind between two elevations, the sense of the difference being taken from the lower to the upper. The ideal thermal wind blows along the isotherms, with colder air to the left in the northern hemisphere. The vertical shear of the actual wind is usually close to the thermal wind. Our understanding of cyclones, anticyclones, fronts, and their associated weather patterns is based on the tendency of the actual winds and the temperature patterns to adjust continuously so as to maintain a close balance between the actual wind shear and the ideal thermal wind.

Trade Winds. The easterlies that blow persistently over much of the oceanic regions of the western hemisphere between latitudes 10° and 25°, and were relied on by the sailing ships of the early era of maritime trade. They usually have a northerly component in the northern hemisphere and a southerly component in the southern hemisphere.

Tropical Cyclone. A weather system that forms over the tropical oceans with a vortex of winds, destructive at times, around a small center of low pressure. The tropical cyclone differs from the cyclones of higher latitudes in that the horizontal contrast of warm and cold air plays no role in its development.

Trough. Referring to pressure, a region of relatively low pressure extending outward from a center, analogous to a valley leading out from a hollow.

Typhoon. A tropical storm in the western North Pacific area, in which the maximum winds are at least 64 knots. Similar storms in other portions of the tropics are called hurricanes.

Veer. A change of wind direction in a clockwise sense. The term is used to specify the change with time at a given point, or the change either with elevation or in the horizontal at a given time.

Vortex Roll. A helical flow of wind in the lowest few thousand feet of elevation, with a horizontal axis usually aligned along the low-level wind. It often has an important effect on the wind at the surface of the sea.

Warm Front. A front, typically extending eastward from a cyclone center, which is moving across the map so that the warmer air is advancing toward the region of colder air. Thus if you are at rest at a given point on the sea surface, you will experience a sudden rise of temperature accompanied by a clockwise wind shift, typically from southeasterly to southwesterly, and perhaps a cessation of the rain occurring to the east of the cyclone, as the warm front passes.

Warm Sector. The sector of a frontal cyclone, usually toward the southeast, in which the air is relatively uniform and warm.

Water Vapor. The gaseous form of water. Its importance to life itself, let alone meteorology, is enormous, despite its small proportion in the mix of gases that constitutes air. By weight this proportion varies from less than one part in 10,000 to as much as about one part in 40.

Waterspout. An intense vortex with an approximately vertical axis, running from the sea surface to a cumulus cloud overhead. It is similar to the tornado over land, but is usually not so powerful and destructive.

Wave Period. The time between wave crests, as measured by a stationary observer.

Weather System. A recurring pattern of wind, pressure, temperature, cloud, and rain, which may cover an area spanning as little as ten miles to as much as a thousand miles. The elements of the pattern have a typical arrangement relative to each other. A cyclone, with wind swirling counterclockwise around a center of low pressure, with warmer air to the southeast and cold to the northwest, and with cloud and rain ahead and over the center, is an example of a weather system. A thunderstorm complex, with a leading squall of strong wind and sharply cooler air followed by a brief but intense deluge of rain, is another example.

Wind Shear. The change of the wind vector with position at a given instant. Vertical wind shear, the change of wind direction and speed along the vertical, is most commonly meant, but lateral wind shear, the change in the horizontal perpendicular to the wind direction, is sometimes discussed as well.

Index

Acapulco, 1986 Olympic yacht races at, 110
Adiabatic temperature change, 16–25, 38, 47, 124. *See also* Lapse rate; Temperature
Admiralty Inlet, in Puget Sound, 109
Advection, 37, 46, 49–50, 51, 126, 132, 133. *See also* Air; Temperature
African Coast, 119
Afternoon low, 130. *See also* Pressure
Air, 9, 11, 20–21; advection of, 46, 51; ascent and descent of, 4, 21, 25, 45, 46, 49, 73, 78, 94, 122, 125; average temperature of, 35; cloudy, 124; column, 117; compression of, 17; convergence of, 96; current, 117; dew point of, 19; discontinuity between cold and warm, 9; displacement of, 24; drift of, 94; elevation of, 119, 124; flow, 86, 87, 91, 92, 117, 124; moisture in saturated, 116; northward acceleration of, 35; parcels of, 15–16, 17, 18, 19; patterns and movements of, 34; smoke laden, 129; stability of, 107; stable or unstable, 20, 33–34, 83, 85, 90, 93, 94, 117, 140; temperature of, 34; tropical origin of, 142; undercutting or overrunning, 9. *See also* Aloft; Cyclones

Air stream, 140; converging, 88; diverging, 88; stability of, 86
Aloft, air, 30, 39, 82; accelerates northward, 35; flow, 99, 132; high temperatures, 74; low pressure, 35; ridges and troughs, 43; seaward, 92; vertical shears, 35, 74; westerly flow, 36
Altitudes, 35
Anemometer, 20, 29, 31, 32, 33–34, 144
Anticyclones, 3, 4, 12–13, 38, 39, 49, 130, 143; behavior of, 43; define, 42; examples of, 45; observed on weather map, 44; over continents, 44; passage of, 140; southern quadrant of an, 90; in upper regions of hurricane, 125
Appalachians, 90
Atlantic Ocean, 116, 117, 119, 122, 123, 140
Atmosphere, 140; at sea-level, 26, 124; balance of, 13–14, 19–20, 23, 24, 36, 46; changing patterns, 9, 20, 29, 36; depth of, 34, 117; geostrophic equilibrium, 47; layers, 45; stability of, 47, 118, 133; thermal wind concept of, 35. *See also* Temperature; Winds
Atmospheric Environment Service (AES), 153
Atmospheric pressure, 16, 130, 144; flow, 46

191